# Parkinson's and Alzheimer's Today

Lars P. Klimaschewski

# Parkinson's and Alzheimer's Today

## About Neurodegeneration and its Therapy

Lars P. Klimaschewski
Institute for Neuroanatomy
Medical University
Innsbruck, Austria

ISBN 978-3-662-66368-4        ISBN 978-3-662-66369-1    (eBook)
https://doi.org/10.1007/978-3-662-66369-1

Responsible Editor: Christine Lerche
This Springer imprint is published by the registered company Springer-Verlag GmbH, DE, part of Springer Nature.
The registered company address is: Heidelberger Platz 3, 14197 Berlin, Germany

*Dedicated to my parents*

# Preface

This book is addressed to all those who want to be informed about the state of research on aging and cell death in the central nervous system. The early development and degeneration of our brain later in life are described with reference to recent research findings. The book is aimed at a readership interested in neurobiology or medicine and deals with the following questions: Why do we need billions of nerve cells? What distinguishes our brain from that of other mammals? Why does only man think in a complex language and plan his actions far into the future? Which parts of the brain are particularly important for this ability? Why do nerve cells fail in old age? What cell biological mechanisms are responsible for this? What exactly happens in Parkinson's and Alzheimer's disease? Can the progression of neuronal cell death be delayed or even stopped? What new therapies for dementia and Parkinson's might be available in the future? Those who are willing to look behind the laboratory doors of neuroscientific research together with the author will receive answers to these questions.

In the first chapter of this book, the basic structure and development of our brain is outlined. The discussion of comparative aspects makes it clear why we need billions of nerve cells in contrast to most animal species. Only with this large number of building blocks, which function as nodes in the multitude of neuronal networks, are the typical human abilities made possible. I will point out the essential differences between humans and other mammals and refer to the neuroanatomical pecularities of our brain that explain how higher cognitive performance is achieved.

In the second part of the book, the consequences of the loss of neurons will be in the foreground. Early on, neurons perish every day (by the age of 80, around a third of all nerve cells in the brain are lost). That this happens largely unnoticed, we owe to a pronounced back-up. The most important information is stored several times in different neuronal networks, so that vital functions usually remain well preserved into old age. For people suffering from a neurodegenerative disease, however, this looks quite different. In addition to the biological basics of aging, this chapter therefore deals with the cellular processes that underlie Parkinson's and Alzheimer's disease in more detail. The focus is on the development of both diseases from a neuropathological point of view, less on the medical side. This book is therefore not intended to replace the medical textbook, but to supplement it with neurobiological aspects.

The third chapter deals with the different options we have to delay or even stop neuronal degeneration. In the future, it may be possible to compensate for lost neurons by enhancing the formation of new nerve cells, the so-called neurogenesis. The currently available therapeutic approaches for the treatment of neuronal degeneration are also presented. However, the focus is on new research results leading to novel therapeutic strategies to treat Parkinson's and Alzheimer's disease.

This book uses work published by leading neuroscientists and physicians worldwide. Their latest work is listed at the end of the chapters (without claiming to be complete). The schematic drawings were created using commercially available templates (https://www.motifolio.com/). What is special about this book is that it will never be finished. Regularly I will present the latest and most relevant developments from the Alzheimer's and Parkinson's laboratories in **Klima's Brain Blog** (https://www.klimasbrainblog.com/en). On my website you may also want to subscribe to a **newsletter** that will keep you up to date.

I would like to express my special thanks to my family and my colleagues and friends for their corrections and comments, in particular to Annegret Wehmeyer, Gerrit Krupski, Dietrich Lorke, Erich Brenner, Christian Humpel, Willi Eisner and Maximilian Freilinger. Furthermore, I would like to thank Dr. Christine Lerche and Claudia Bauer from Springer-Verlag for their support. Finally, I would like to thank all the students of the Universities of Heidelberg and Innsbruck who have attended my lectures and seminars over the years and discussed the development and aging of the brain with me. Some of the aspects discussed in this book go back to these conversations.

Most of the conclusions reported here are based on generally accepted and confirmed scientific results, however, some questions remain open and require further investigation. But that's how science works. Elaborate experiments are carried out in the laboratory and large data sets are generated. With much skepticism, these are then checked and often discarded, because the following sentence by Charles Darwin, the discoverer of the theory of evolution, applies more than ever: "False facts are extremely harmful to the progress of science, because they often persist; false theories, on the other hand, which are supported by some evidence, do no harm; because everyone strives with praiseworthy zeal to prove their incorrectness".

# Contents

# 1

# Introduction to Brain Development: Why do We Need so Many Nerve Cells?

Our brain belongs to the central nervous system (CNS), which also includes the spinal cord, and can be divided into three parts: cerebrum, brainstem (truncus cerebri) and cerebellum. The latter hangs at the back of the brainstem, which comprises the medulla oblongata, pons and mesencephalon. In the cerebellar cortex most of our nerve cells (neurons) in the CNS are located. Of the approximately 90 billion nerve cells in the human brain, 60–70 billion are found in the cerebellum, although it makes up only 10% of the total brain mass of about 1400 g (it is about 140 g heavier in men than in women). Despite the undoubtedly demonstrable differences in female and male brain development, the blueprint of the brain is largely the same between the sexes and the absolute brain weight is of secondary importance (as we will see below).

The forebrain, the prosencephalon, consists of the diencephalon and the telencephalon. The latter comprises four large outer lobes (frontal, parietal, occipital and temporal lobes) and a smaller insular lobe, which lies behind the lateral sulcus, a long and deep furrow (Fig. 1.1). The lobes carry the cortex cerebri with a total of about 16 billion neurons. Underneath we find the basal ganglia—basal because they are located deep below the cortex—which are particularly important for understanding Parkinson's disease.

Furthermore, the limbic system is of importance, in particular for our spatial and declarative memory, i.e. memory bound to words and numbers.

© The Author(s), under exclusive license to Springer-Verlag GmbH, DE, part of Springer Nature 2022
L. P. Klimaschewski, *Parkinson's and Alzheimer's Today*,
https://doi.org/10.1007/978-3-662-66369-1_1

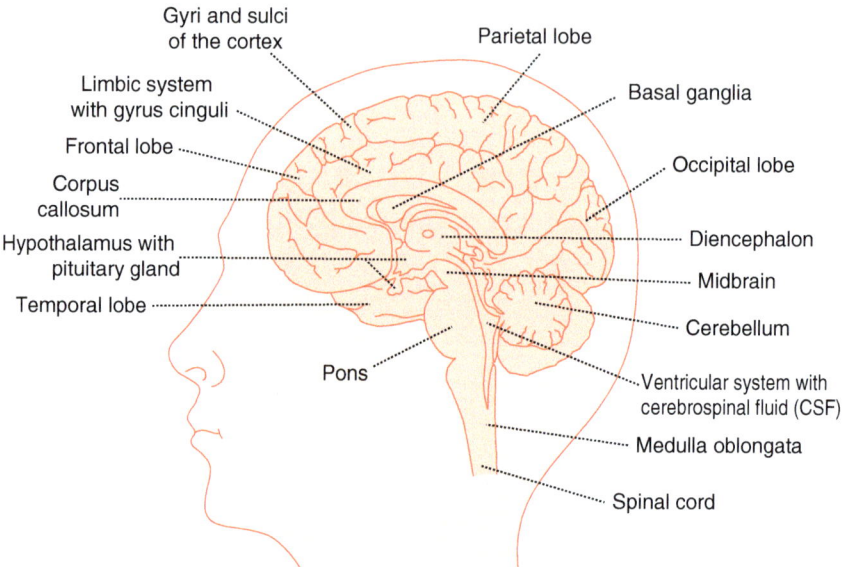

**Fig. 1.1** Shown in this schema are the frontal, parietal, temporal and occipital lobes of the brain, divided in the midline. The insula (Insula), the fifth lobe in the brain, can only be seen from the lateral side. The brainstem includes the medulla oblongata, pons and midbrain. The diencephalon is found above the midbrain, deep in our brain. The cerebellum is located dorsally on the brainstem in the posterior cranial fossa. Inside the brain there are several cavities, the ventricular system, which are filled with cerebrospinal fluid

It comprises large parts of the temporal lobe and the Gyrus cinguli, which is located above the largest connection of the two hemispheres, the corpus callosum. The failure of these structures, for example in Alzheimer's disease, affects, therefore, both the semantic memory (factual knowledge) and the retrieval of personal experiences, the episodic memory.

The embryological and fetal development of our brain from the beginning of pregnancy to birth is a highly organized and genetically largely determined process, which will be considerably influenced by the environment after birth. However, these postnatal changes at the morphological level are often only visible after magnification with a microscope. We assume that the fine-tuning of the connections between neurons during development continues beyond the 20th year of life, especially in the frontal brain regions located behind the forehead. However, the formation and removal of contact points (synapses) takes place throughout life.

## 1.1   Neurons and Glia in the Central Nervous System

The stem cells of the nervous tissue not only develop into nerve cells, the neurons, but also into glial cells. Neurons have a single axon as an efferent extension leading away from the cell body to transmit information, and usually several afferent dendrites, cellular processes that receive information. For each neuron, there is approximately one glial cell in the brain, but the ratio varies widely by brain region. The glia takes up the space between nerve cells and constantly interacts with them at the cellular and molecular level, so astrocytes and oligodendrocytes, the macroglia, are very important for the overall functionality of the brain (Fig. 1.2).

The most important task of oligodendrocytes is the formation of the myelin sheath around the axons of the nerve cells, called myelination. Similar to

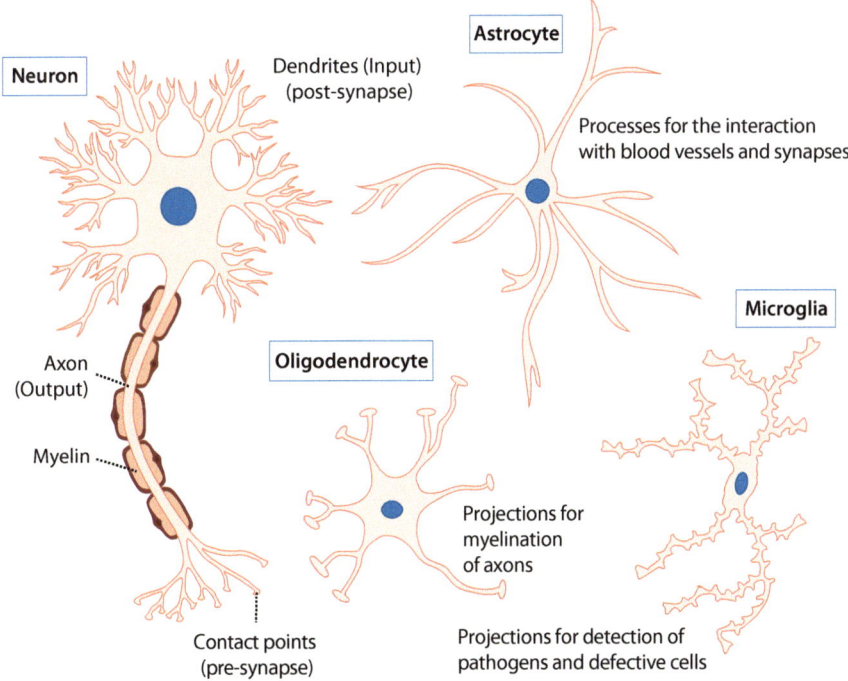

**Fig. 1.2**   In addition to the nerve cells (neurons) that have an axon and multiple dendrites, there are essentially two types of macroglia (astrocytes and oligodendrocytes) as well as the microglia, which acts as the brain-resident defense system. Glial cells also have cytoplasmic processes that allow them to radiate into the environment and establish contacts with neighboring cells

the electronics, the axons are electrically insulated so that no short circuits occur and the action potentials can be transferred faster from one nerve cell to the next. Oligodendrocytes have several cellular extensions that myelinate up to 40 axons at the same time. This allows conduction speeds of 200 m/s to be achieved in vertebrates. If there were no myelin, individual axons would have to be up to 1 mm thick to achieve such high speeds of conduction. However, their diameter is only a few micrometers.

Another type of glia is formed by the astrocytes, which are mainly responsible for maintaining the metabolic balance of the extracellular space, i.e. they control the concentration of important molecules (ions, transmitters, etc.) in the cerebrospinal fluid. The extracellular fluid of the CNS is produced in the interior of the brain, the ventricular system, by a capillary network, the choroid plexus. In addition, astrocytes supply the nerve cells with nutrients and surround the perivascular space around the blood vessels. In this Virchow-Robin space cerebrospinal fluid flows, which—similar to the lymph in our body—transports cellular breakdown products and non-reusable proteins out of the brain. Finally, astrocytes surround the synapses, so that the released neurotransmitters will not diffuse into the environment, but remain between two nerve cells.

In addition to the two types of macroglia, there is the microglia, which plays an important role in brain injury and in various neurological diseases by taking over the function of immune cells in the CNS (Fig. 1.2). In the brain and spinal cord, only very few white blood cells (lymphocytes) can be found, i.e. the brain is extremely sensitive to germs and viruses and inflammations caused by them.

Post mortem (after death) anatomical and histological examinations of the brain, but also high-resolution imaging methods in the living brain have given us a good insight into the complexity of the brain and significantly enhanced our knowledge of neurological and psychiatric diseases. Moreover, the myelination can also be examined more closely, as it may be altered in autism or schizophrenia.

## 1.2    What Happens During Brain Development?

The formation of myelin sheaths in the area of the long tracts occurs in the CNS until well into the second year of life. Only then the isolation of the axonal fibers, for example in the lower part of the spinal cord, is completed.

We cannot expect a conscious, voluntary control of our sphincters at the pelvic floor in babies (potty training for toddlers makes no sense at all in the first year of life). In the brain, the formation of myelin sheaths starts late and, as explained below, continues in some areas until the second decade of life.

The external shape of the brain is already well recognizable at the end of human embryonic development, four months after fertilization of the egg. Five brain vesicles can be distinguished. They form the structures of the telencephalon, the diencephalon, the midbrain, the pons, and the medulla oblongata. The latter two structures together with the cerebellum form the rhombencephalon. Each of the two telencephalic hemispheres are formed from the lateral and upper wall of the forebrain structure, while the diencephalon develops from the inner, medial wall. The eye and the thalamus represent diencephalic structures. The hypothalamus, which controls our vegetative, autonomic nervous system and endocrine organs, develops from the thalamus. The pituitary gland hangs from the hypothalamus and is anatomically and functionally closely related to it (Fig. 1.1).

The nerve cells in the cortex and in the subcortical nuclei arise from stem cells, which are located in the vesicle wall, the ventricular zone. After division, their daughter cells migrate as neuronal progenitors along the processes of a special type of glia, the radial glia, into the outer mantle of the pallium, the future cortex cerebri. We speak of the neocortex when the neurons come to lie in six superimposed layers (laminae). In the pallium, the cells are unordered and still have to migrate to their final location. So, at the end of development, the cortex consists of six layers (laminae I-VI, Fig. 1.3). In layer III and V the large nerve cell bodies with long axons are located (pyramidal cells). In layers II and IV are rather small neurons, whose axons are short (granular cells).

In the third month of development, the pallium grows in several directions, one forward (frontal), one backward (occipital) and one downward and outward (temporal), so that some parts of the cerebral cortex as well as several underlying structures reveal a bow-shaped form that resembles a ram's horn. The cerebral hemispheres consist for the most part of neocortex, which in human embryogenesis almost completely covers our brainstem. Developmentally, there are two older parts, the palaeocortex and the archicortex, which remain visible in the lower and medial parts of each hemisphere. In the early mammals that lived 100–200 million years ago and in today's dogs, cats or cows, these phylogenetically oldest parts of the cortex are significantly larger in relation to the neocortex.

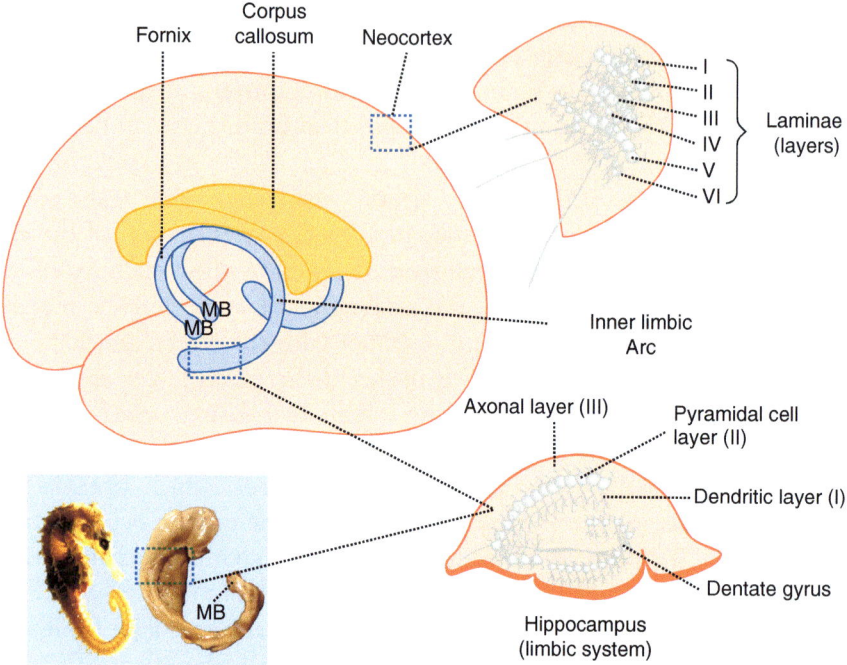

**Fig. 1.3** The limbic system, which forms arches around the diencephalon, lies deep in the brain (the word limbus means "border"). The fornix indicates an important fiber tract projecting into the mamillary body, MB). The corpus callosum, the largest axonal commissure connecting the two hemispheres, is located above the fornix. The neocortex is evolutionarily younger and more complex than the archicortex or paleocortex. The hippocampus resembles a seahorse and consists of a three-layered archicortex. The paleocortex, the phylogenetically oldest part of the cortex, which is two- to three-layered, is located at the base of the brain in extension of the olfactory bulb, which is part of the first cranial nerve (seahorse by Laszlo Seress, CC-BY-SA)

## 1.3    Evolutionarily Old Brain Parts are Simpler in Structure than the Neocortex

The palaeocortex forms the olfactory brain and is rudimentarily present in humans only. The archicortex, on the other hand, is mainly found in the hippocampus of the temporal lobe and plays an outstanding role in the formation of emotions and memory. In the animal kingdom, it is vital to be able to remember an attack by a predator to adapt the future behaviour accordingly. On the other hand, successful hunting grounds must also be stored. These functions are implemented by the hippocampus, the central structure of the limbic system, in cooperation with the neocortex.

The hippocampus looks like a lying seahorse (hence the name) and is discussed in more detail in Chap. 2 in the context of Alzheimer's disease. It is a structure located at the floor of the lateral ventricle in the temporal lobe and consists of a three-layered archicortex (Fig. 1.3). The dentate gyrus forms a layer of small neurons (granule cells) next to an outer dendritic layer and an inner layer with axons that innervate the dendrites of the pyramidal cells in Ammon's horn of the hippocampus (cornu amonis, CA). The densely packed neuronal cell bodies (perikarya) of the CA form the middle layer (II). Dendrites are in layer I and III, but the axons only in layer III. Some smaller cell bodies, the interneurons, are found as well. The axons of the pyramidal neurons form the inner layer as "fimbria" in a long and curved path, named the fornix that projects to the corpora mammillaria (mammillary bodies, MB in Fig. 1.3).

The stem cells lying at the bottom of the hemispheric vesicle give rise to the early neurons of the ganglionic eminences. From this, the later basal ganglia (nucleus caudatus, putamen, globus pallidus) arise. They form the largest nuclei in the depths of the brain and are primarily involved in the regulation of motor (movement) programs. Moreover, they control complex psychomotor behavior, learning processes, and participate in cognition and emotions.

The basal ganglia and the cerebral cortex are controlled by phylogenetically very old nuclei in the brainstem, i.e. those found already in the earliest vertebrates. Since these grow much slower over the millions of years than the structures of the forebrain, the axonal processes of the nerve cells located in the midbrain, pons or medulla oblongata have to branch out extensively to innervate all their target neurons in the brain. This increasing morphological complexity poses very high demands on neuronal metabolism and makes the neurons more susceptible to cellular stress. They therefore fail first in old age and in neurodegenerative disorders, e.g. in Parkinson's disease.

**In a Nutshell**

- The human brain contains 80–90 billion nerve cells, 60–70 billion of which are in the cerebellum and 16–20 billion in the cortex, which consists of paleo-, archi- and neocortex.
- The archicortex forms the hippocampus that plays an important role in our emotions and memory processing.
- In the hippocampus formation, new nerve cells are being formed in the first 20 years of life. This process is called neurogenesis but decreases significantly and is hardly detectable in old age in humans.

- The basal ganglia, located below the cortex, regulate and modify most of our motor activities.
- The cortex and the basal ganglia are controlled by neurons in the brain stem, which form highly branched axonal trees to innervate a large number of neurons in the phylogenetically younger forebrain areas.

## 1.4    What Distinguishes the Left from the Right Brain?

Functionally, the two hemispheres are different, one speaks of hemispheric asymmetry or lateralization. The hemispheres communicate with each other via commissures, of which the corpus callosum is the largest connection. It couples homologous cortical areas via the projections of the large pyramidal cells that project with their long axons to the opposite side. This makes most cortical regions directly connected with its partner region in the other hemisphere and allows the exchange of information, e.g. on complex movement programs before a motor activity is sent to the brainstem or to spinal cord neurons, which then activate our muscles.

The left hemisphere is particularly responsible for the formation of language (especially syntax, i.e. sentence structure and grammar), but also for reading, writing, arithmetic and language-related memory. The left side processes information sequentially, similarly to the processor in a computer, and tries to establish cause-effect relationships. It works analytically, for example, by breaking down a larger mathematical problem into manageable, smaller tasks. With lesions of the left hemisphere, therefore, problems in analytical thinking and especially in language capabilities occur.

In contrast, the right hemisphere looks for analogies and similarity relations. It works holistically and is integrative, e.g. a complex spatial structure consisting of many individual parts is recognized as a whole. The right hemisphere also helps us to discern faces, to understand language in general and to form the sentence melody (prosody). Furthermore, it is necessary for the non-verbal memory and for the sense of direction. Lesions in the right parietal lobe are characterized by disturbances of the spatial-coordinative abilities.

Interestingly, the language dominance of the left hemisphere is not always coupled with right-handedness. The right hand is indeed controlled by motor areas located in the left hemisphere, i.e. most connections in our brain are crossed. But the language center is usually on the left in left-handers as well. We still do not know exactly how lateralization, i.e. the dominance by one of the two hemispheres, arises. Since it is not very pronounced in our closest relatives, the apes, the establishment of a connection between the right ear (the preferred input for language) and the left hemisphere is generally attributed to the upright walk of man. This is explained by the fact that the unborn child usually lies in the womb so that the right body and facial side face outwards. This stimulates the right ear during the mother's walking and speaking in the development. It thus forwards more signals to the brain than the left ear and thus leads to an increased use of the left hemisphere even before birth. This would explain the right-handedness in most people.

## 1.5 Brain Development in Childhood and Adolescence

Imaging studies in children have shown that the cerebral cortex enlarges especially in the frontal and parietal lobes from the 10th to the 12th year of life. An increase in volume of the cortex can even be observed in 16-year-old adolescents with respect to the temporal lobe. Differences between the sexes suggest a role of sexual hormones in the morphological changes that occur during puberty. Particularly in the frontal lobe, neuroplasticity, the formation and degradation of synapses, is still pronounced at this time. Hence, the highest level in the hierarchy of the CNS, the neocortex, undergoes long-term changes after birth that continue well into the second decade of life.

In the hippocampus, the formation of new nerve cells is ongoing during the first ten years of life. This postnatal neurogenesis is possibly of great relevance to the development of new therapies for Alzheimer's dementia and other neurodegenerative diseases. However, it is still not clear what the newly produced neurons in the human brain are actually needed for in children (more on this in the third chapter). So far, the continuous division of nerve cells has been analyzed carefully only in some animal species, for example in singing birds, which learn a new song after new neurons have been generated in the vocal center of their brain.

In addition to neuronal cell division and the outgrowth of cellular processes, the formation of myelin sheaths can also be observed for a long time after birth in some brain regions. We assume that myelination in the frontal lobe is not complete until around the 30th year of life. At this time, the construction of the "hardware" in our brain would finally have finished, and we could expect the greatest cognitive brain performance at this age. In fact, many Nobel laureates (at least in the natural sciences) made their prize-worthy discoveries at this relatively young age.

Both, nerve cells and synapses, are created in excess before birth. Most of the surplus contacts between neurons disappear in the first decade of life, leading to reduced cortical activity. So, up to the age of 5 we have the highest synaptic density in the cortex. In the prefrontal cortex (located directly behind the forehead and lateral to the eyes), the density of synapses decreases up to the age of 20. We then have the enormous number of approximately 1000 trillion synapses in our brain. Probably many of the initially formed contacts are not part of functionally relevant neuronal networks. They therefore remain unused and may in part be eliminated by microglia and astrocytes.

## 1.6    The Child's Brain is Enormously Plastic and can Still Heal

As already mentioned, cultural influences play a significant role in the fine-tuning of neuronal connections after birth, whereas the long axonal projections between the cortex, the subcortical nuclei and the spinal cord are genetically determined. In contrast, neuroplasticity remains high throughout life particularly in the hippocampus and in the neocortex, but in limited areas of a few dozen micrometers in diameter only.

Sensory input drives the formation of new synapses. On the other hand, less used synaptic contacts must be removed. Otherwise our brain would continue to increase in volume and weight over time, which would not be possible due to the limited skull volume. The total number of synapses between neurons therefore does not change significantly in the healthy brain after development is completed and before the aging processes set in. This ensures the synaptic homeostasis, the balance of newly formed and eliminated synapses.

In line with the high activity and plasticity up to approximately the fifth year of life, a pronounced ability to regenerate is observed in children's brains even after severe brain injuries. This is possibly due to the re-activation of superfluous synapses that have not yet been removed. For example, if the entire left hemisphere is lost in the first years of life, the language ability may be taken over by the right half of the brain. Also, sensory areas that are normally responsible for the processing of visual and auditory impulses are not inactivated in children who are born blind or deaf, respectively, but are extensively used to process the remaining, intact sensory information, e.g. those originating from tactile receptors (mechanosensation). These children then become significantly better in the perception of pressure and touch sensations than children who are able to see and hear.

The cortical size of a given sensory area and thus the number of nerve cells available for a particular function determine the performance of the brain. Children who are born blind often have "fine ears" and, conversely, deaf children may process visual impressions very well, so they have "good eyes". The respective brain regions that would have been responsible for the failed functions are taken over by other, intact sensory systems. However, this impressive plasticity is limited to the first few years of life and can only be reactivated to a very limited extent later on.

**In a Nutshell**

- The development of our brain up to birth is a highly organized and genetically largely determined process, which is subsequently influenced by the environment.
- Many structural changes still take place in the brain during the first years of life, which may be easily disturbed. From the fourth year of life (kindergarten age) the brain becomes more resistant to psychosocial stressors.
- Nerve cells and synaptic contacts are laid down prenatally in great excess. Most of the superfluous synapses disappear during the first decade of life.
- It is not until the fourth decade that the myelin formation in the frontal brain is finally completed.
- The child's brain can still heal after injuries because of it's much higher plasticity than the brain of adolescents and adults.

## 1.7 Is a Large Brain "Smarter" than a Small One?

The size of the brain was often taken as a substitute for the functional capacity and thus the mental performance in comparative and cognitive neuroscience. As mentioned above, the brain weight in humans is on average 1.4 kg. On the other hand, there are vertebrates whose brain weighs only 1 mg. The brain of a mouse is around 1 g light (Fig. 1.4). The brains of elephants weigh about 5 kg and even up to 10 kg for blue whales. Does this now tell us that elephants are five times and whales ten times as smart as we are? Obviously this is not the case, because it is we humans who determine the fate of elephants and whales and not vice versa. However, almost all animals have

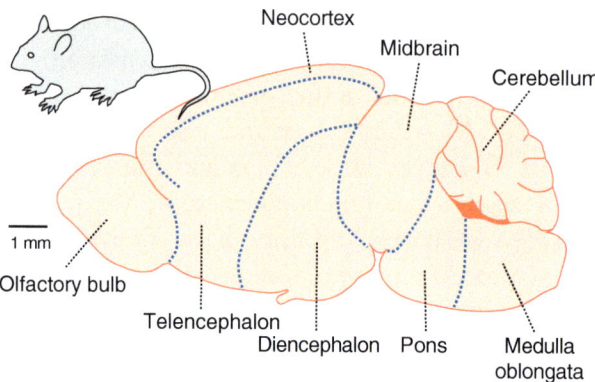

Neocortex

Midbrain

Cerebellum

1 mm

Olfactory bulb

Telencephalon

Diencephalon  Pons

Medulla
oblongata

**Fig. 1.4** Approximately 60–80 million years ago, the common ancestor of mouse and human lived. The mouse brain is structurally similar to the human brain, but has a relatively bigger olfactory bulb and a smaller neocortex without gyri or sulci. The brainstem is significantly larger than the forebrain (note the scale bar on the left) when compared to humans

some skills in which they are far superior to us humans. Examples would be the much faster running cheetah or the bird of prey that observes a mouse over long distances. There are also snakes that can see very well in the dark due to special receptors in their heads for long-wavelength (infrared) light. This makes it possible to perceive prey animals that emit heat, such as rabbits, also at night.

The special abilities of animals encoded in various sensory and motor parts of the brain apparently are not sufficient to ensure their survival against attacks by other species, even if they are bigger and have heavier brains. A central message of comparative neuroanatomy is, therefore, that it is not the absolute brain weight, but the absolute number of nerve cells accomodated especially in the cortex cerebri which decides the cognitive performance and thus ultimately the survival of an organism. African elephants have very large brains and in total three times as many nerve cells as humans, but their cortex is much simpler structured and contains fewer neurons (6 versus 16 billions, see Table 1.1).

For whales, the story is more complicated. The pilot whale (a dolphin species) has significantly more neurons in the cortex than humans (over 30 billion). However, the dolphin cortex only allows for a maximum of five distinct layers (laminae). Also, the neurons of dolphins are less strongly interconnected, which is reflected in the smaller thickness of the cortex (1–2 mm) in comparison to humans (2–4 mm). Therefore, humans have the widest cortex, independent of brain size, in order to accommodate and

**Table 1.1** Comparison of important quantitative parameters between the brains of elephants, dolphins, humans and rats. The elephant has the heaviest brain, the dolphin the largest surface area and the highest absolute number of nerve cells in the cortex. Interestingly, the density of neurons in the neocortex is highest in a small rodent, e.g. in the rat

|  | Elephant | Dolphin | Human | Rat |
|---|---|---|---|---|
| **Brain weight (g)** | **4800** | 3600 | 1400 | 2 |
| **Neocortex (mm²)** | 260000 | **375000** | 230000 | 600 |
| **Neuron density (/ mm³)** | 5000 | 20000 | 24000 | **55000** |
| **Number of neurons (x 10⁹)** | 6 | **37** | 16 | 0,03 |
| **Thickness (mm)** | 2-3 | 1-2 | **2-4** | 1-2 |

interconnect as many neurons as possible. When comparing different animal species in terms of cognitive brain performance, we must take into account not only the weight of the brain, but also the histological and cellular features.

Large marine mammals perform apparently poorer with over 30 billion nerve cells in the cortex than we humans or the apes do with less than half of this number of neurons. Many primates score significantly better in species-specific cognitive tests than dolphins, although the latter have more neurons. It is not clear why the number of neurons in certain whale species (cetacea) increased so strongly during evolution. It may be due to the fact that marine mammals have dived into ever deeper and colder areas of the sea over the millions of years and had to generate more heat to survive in these depths. In fact, there are indications of a high synthesis of heat-producing proteins in the mitochondria, the cellular "power plants" in the brain of dolphins. That would be an evolutionary advantage that had nothing to do with cognitive abilities.

During the evolution of species, the large, late-developed brain regions became increasingly structured. This resulted in layering of the neocortex into the aforementioned six layers (laminae) and allowed groups of cortical nerve cells referred to as modules to be controlled more precisely than it would be the case in a spatial jumble. Neurons became interconnected with each other on the shortest possible paths, while the number of synaptic connections per neuron remained approximately the same on average. So it did not increase proportionally (not every neuron is connected to every other one). Otherwise our head would be oversized. The synaptic density in the cortex of hedgehogs and monkeys is approximately the same, even though the cortex volume is over 100 times larger in the latter. These observations underscore the importance of increases in the absolute number of neurons during phylogenetic development as an indicator of higher cognitive performance.

## 1.8   Absolute and Relative Brain Weight

In this context, the relative brain weight is often discussed (Fig. 1.5). It must not be overlooked that a blue whale has to control a body weighing up to 100 tons with its brain of 10 kg. Therefore, its brain makes up only 0.01% of its body weight, while the relative brain weight (i.e. the proportion of the body weight) in humans is 1–2%. Interestingly, the relative brain weight

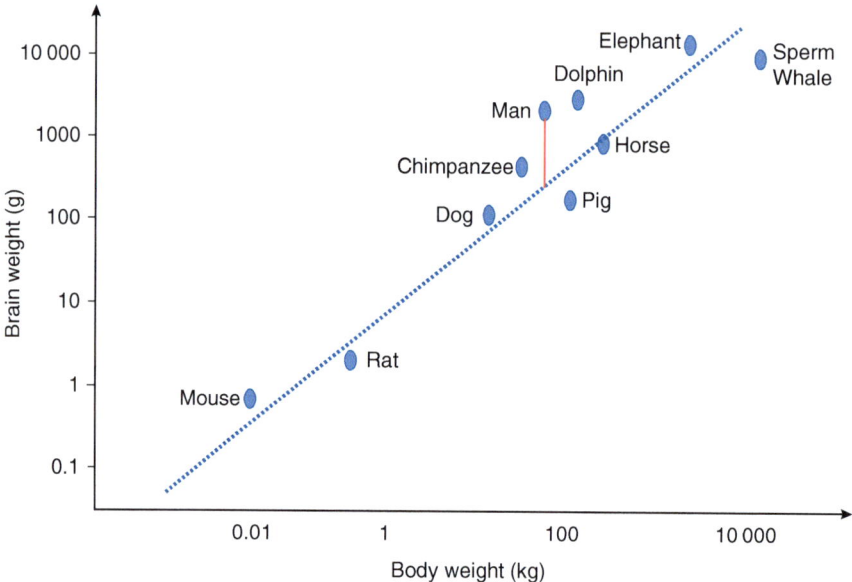

**Fig. 1.5** The brains of mammals scale allometrically, i.e. the increase in brain weight compared to body weight follows a line in a double logarithmic representation. The basis for this is the exponential increase in organ weight through biological cell division (2, 4, 8, 16, 32, etc.). The average brain weight of humans is the farthest above the regression line (dotted). It is significantly larger than one would expect for a mammal of the same body weight. This relative brain weight of humans is not exceeded by any other large animal, only some dolphin and porpoise species are also found far above this best-fit line. In addition to the relative brain weight, the absolute thickness of the human neocortex is the highest in the entire animal kingdom (according to Nieuwenhuys, The Central Nervous System of Vertebrates, Springer, 1998)

of the dwarf mouse, one of the smallest rodents at all, is still significantly higher with 4% of its body weight. In elephants and horses it is lower, however (0.2%). Chimpanzees, which are genetically closest to humans (more than 95% identical genome), have a brain 3–4 times smaller than ours and a relative brain weight of 0.9%.

Overall, the comparison of mammals with each other shows that our brain weight is 7–8 times higher than expected for any other mammal of the same body weight. This factor is expressed by the encephalization quotient (EQ), which is highest in humans compared to all other animal species. Although apes such as gorillas and orangutans are on average much heavier than humans, they have a brain that is appr. three times lighter.

As the example of the dwarf mouse shows, a correlation of EQ with cognitive performance is not always given, however. Rather, the relative brain weight must be considered in relation to the absolute number of nerve cells in the cortex in order to be able to compare cognitive abilities of different species with each other. Here, in particular, a correlation is seen with the ability to form complex social structures. Interestingly, the two parameters (weight and cell number) are only weakly linked to each other with respect to the different orders within the mammal family: While a ten fold increase in the number of neurons in rodents is associated with a 35-fold larger brain, primate brains with ten times more neurons are only 11 times larger. This saves costs, because the upright posture is not made more difficult by the relatively lighter head. In addition, the head still fits through the birth canal in the female pelvis. The reverse comparison is also interesting: Analogous to the human with its 86 billion nerve cells in the brain, the brain of a rodent would have to weigh about 35 kg if it would contain the same number of neurons!

So our brain has increased significantly in size in relation to body mass during the past 2–3 million years of evolution, not in a steady manner, but in two big steps. The first increase happened about 1.8 million years ago, followed by a second rise about 300,000 years ago. Although the great discoverer of biological evolution, Charles Darwin, assumed a slow, linear development of these parameters, we now know that many adaptations occurred abruptly, and only few structural changes took place in between, i.e. over hundreds of thousands of years. This fact is known as punctuated equilibrium.

The first growth spurt required the brain to use significantly more energy, which suggests that the increase in its size is associated with the beginning of hunting animals by the upright walking human (Homo erectus). Already 2.5 million years ago, early humans in Africa began to make tools from stone, such as rasps and axes. This made it possible to switch the main energy supply from a plant-based diet to a protein-rich diet. The intake of fried meat with high energy density made it possible for the body and the brain to grow much faster than would have been possible with a purely plant-based diet.

In this context, it is important to know that, although the human brain makes up only about 1–2% of our body weight, it uses 20% of the total energy consumed by food. The energy flowing into the brain corresponds to about 15 watts, which is about as much as a weak light bulb. Of all other organs, only the gastrointestinal tract is metabolically more intensive. Therefore, over the course of development from early humans (Australopithecus) to Homo sapiens, the blood supply of the brain had to increase significantly as well. In fact, an enlargement of the brain-supplying arteries can be observed over the years, as imprints of vessels are detected on fossilized skull bones. Their passage points through the base of the skull can be measured. In modern humans, about 10 ml of blood is transported into the brain per second. More than two thirds of the energy carriers (in particular sugar) taken up are needed for the production of neurotransmitters and their release machinery at the synapses. Interestingly, the blood supply to our brain is relatively constant, regardless of whether we are awake or asleep, exercising or solving difficult mathematics problems.

This advantage of a larger and, therefore, more powerful brain, achieved particularly through the change in diet, was used by Homo erectus to make ever more complex tools. For example, improved spears could kill larger animals. The change in food led to an increase in brain size and the increased intelligence in turn led to more effective hunting. Such mutually dependent developments represent a positive feedback mechanism in evolution, which made significant changes possible within a few thousand years. It was probably the major driving force in the development from Homo erectus to Homo sapiens.

## 1.9    With the Second Evolutionary Leap, Our Brain Reaches its Maximum Size

Another significant increase in our body and brain weight can be observed around 300,000 years ago, which then plateaued 100,000 years ago (brain and body size actually starts decreasing after this date). An obvious explanation for the end of brain size growth is a mechanical obstacle: at the end of pregnancy, a baby's head would no longer fit through the female pelvis if the brain and skull continued to grow. One solution to this problem would be to allow the pelvis to grow along with the brain, making it larger. However, calculations showed that an oversized pelvis would severely restrict a woman's mobility, making it difficult for her to escape quickly in an emergency.

In addition to the size of the female birth canal, the maximum performance capacity of the placenta also plays an important role in the early birth date of humans, as a longer pregnancy (from 10 months on) leads to under-supply of the fetus. The human baby is therefore born still immature. It is completely dependent on parental care. Except for independent breathing and swallowing, we can't do much at birth. Therefore, considerable biochemical and structural changes take place during the first years of life in our brain, which can easily be disturbed by negative influences from the outside. It is not until the fourth year of life that our the brain becomes more resistant to stressors.

By the age of six, a child's brain already weighs approx. 90% of an adult brain. The volume increase necessary for autonomous survival is therefore mainly postnatal. In fact, the human brain weight triples during childhood. This enlargement is not due to the formation of new neurons, but rather due to the significant increase in glia and formation of myelin sheaths for faster electrical impulse conduction.

When comparing different mammals, the neuronal density in the cortex decreases much more slowly than the brain weight increases. Therefore, unlike ontogenetic development, the increase in brain size in adults during phylogeny must be due to an increase in the number of nerve cells (the size of each individual nerve cell differs only slightly between most mammals). In order to accommodate billions of cells on an area of about a quarter square meter in the head, the cortex needs to fold. This folding (gyrification) of the neocortex is observed in brains weighing more than 10 g and allows a pronounced surface enlargement without the cortex becoming thicker or the skull larger as a whole.

The increase in body and brain weight 200,000–300,000 years ago, around the time when Homo sapiens appeared on Earth, may have been due to the fact that larger individuals were preferred in mate choice. This would result in the brains of the children from these relationships becoming larger on average as well, which leads to an increase in brain weight quickly (a positive feedback loop). The competition for the largest and strongest partners (intraspecific competition) is likely to have caused the exponential increase in the size of our brains 300,000 years ago.

**In a Nutshell**

- Our brain has increased in size significantly over the course of evolution.
- This increase in weight was not steady. The first increase occurred about 1.8 million years ago, followed by a second one about 300,000 years ago.
- Our brain weight is 7–8 times higher than what would be expected for any other mammal with a comparable body weight.
- Humans do not have the largest brain of all mammals, but the widest neocortex in order to accommodate as many neurons as possible and to connect them with each other.
- Although the brain only makes up 2% of our body weight, it uses 20% of the energy taken in through food.

## 1.10  Neural Stem Cells Remain Capable of Dividing for a Long Time

How does nature actually manage to prolong our embryonic and postnatal development in comparison to other mammals? The most important difference is that human neural stem cells divide more often than those in other species. This results in a significant increase in the number of progenitor cells arising from the stem cells. Thus, the brain regions that develop later in embryogenesis are formed from a larger pool of cells than the brain regions that are completed earlier. The former will be much larger than the latter (according to a rule described by Finlay and Darlington *late equals large*). As a result, the neocortex practically covers all other parts of the brainstem that were laid down first in phylogenetic development.

The intrinsic factors that have allowed us to build a larger brain than other mammals and produce a greater number of neurons in our cerebral cortex are genes residing in the stem cells. When the apes (gorilla, orangutan, chimpanzee) separated from us humans (hominins), a gene duplication (ARHGAP11B) appeared, which significantly increased the division rate of precursor cells and thus led to an expansion, in particular of the neocortex. This is a good example of the importance of gene duplications in the evolution of mankind. Thereby, a new function is assigned to a duplicate of an existing gene. The driving force is, therefore, not the mutation of a gene, which usually impairs the function of the resulting protein, but rather the multiplication of existing genes to increase the variety of proteins necessary to regulate cellular proliferation, differentiation and migration, among others. With regard to the nervous system, other decisive changes in the development of neuronal extensions (axons and dendrites) and in neurotransmission played a role as well. It is estimated that 40–50% of our cognitive intelligence is genetically determined, and the rest acquired post-natally through learning and environmental (cultural) influences.

For understanding of brain diseases which I will discuss in the next chapter, it is particularly important that the neuronal progenitor cells are practically capable of division during development only. Evolution has not produced any mechanisms which would allow for the re-activation of stem cells in the brain (or heart) to such an extent that nerve or heart muscle cells could be replaced in the adult. The reason for this is not yet fully understood, because it would obviously be an advantage if lost tissue could be quickly rebuilt in all of our organs.

One possible explanation for this conundrum is that, after completion of brain development, newly formed neurons in the human brain no longer have a chance of finding the correct target cells from among the billions of possible partners (this would, however, not be a problem for the heart). With regard to the brain, some researchers assume that the loss of regenerative mechanisms in humans was evolutionarily meaningful in order to reduce the risk of incorrect neuronal connections in the adult brain. If the neurogenesis in the human brain were not stopped after childhood, elementary functions and learned programs could be significantly impaired. This hypothesis explains why, in the evolution of primates, genes inhibiting neuronal regeneration were positively selected for, since they apparently were advantageous for the long-term survival of most individuals.

## 1.11 The Frontal Lobe is Especially Important for Higher Brain Functions

The prolonged human embryogenesis and the increase in the proliferation rate of neuronal precursor cells result in a large number of neurons, which allow us to develop specific human behaviors. Over the millions of years the anatomical conditions were created so that only 50,000 to 80,000 years ago an elaborate and highly complex symbolic language could emerge, which only exists in this form in humans. In the frontal lobe, large cortical areas regulate the activity of our facial and speech muscles as well as arm and hand muscles, which together control the verbal but also the non-verbal communication skills that lie at the heart of our social interactions.

This phylogenetically relatively young development has been deduced from datable cave paintings and discovered body jewelry that mark the approximate beginning of symbolization, that is, the doubling of objects by sounds. The possibility of long-term storage of words and numbers as explicit memory plays an important role as well. This step in cognitive development was the decisive one: A conscious world of thoughts was created beyond the world of feelings, since the objects could not only be seen and thought, but also named and thus communicated. The learning through **listening** began, which replaced or supplemented the previously prevailing **watching** as the primary form of information transfer. We know from some patients who lose the ability to communicate through loss of speech or paralysis of their seventh cranial nerve that controls facial expressions, that they would hypothetically be willing to accept blindness in exchange for the restoration of hearing functions.

Verbal communication as the basis of learning complex relationships and profitable teamwork justify a special position of humans in the animal kingdom. The possibilities of direct interactions among members of other species are much less pronounced, although their brains have experienced a significant volume enlargement and re-organization during the millions of years, too. However, the difference to us humans is in the number of neurons and their synaptic connectivity, especially in the neocortex, which covers almost the whole brain and makes up around 80% of the brain weight, although it only contains 20% of all neurons.

However, this ratio alone does not justify our special position in the world. As discussed above, it is similar in some other mammals, and humans have approximately the same overall number of neurons in the CNS which would be expected for a primate brain of the same body size. Therefore, no higher power of any kind, no god or "intelligent designer" is needed to distinguish humans from animals. It is Darwin's theory of evolution which is completely sufficient to understand the making of our brain and, with it, our cognitive abilities as a whole. Unfortunately, evolution has not produced a mechanism yet which protects us from the loss of neurons in old age or in neurodegenerative diseases. This has much to do with the fact that, in evolution, it is primarily about reproduction and not about the avoidance of diseases beyond the reproductive age. I will go into more detail about this in the second chapter.

The final paragraph of this chapter deals with our ability to communicate that is determined to a large extent by the frontal lobe, in which the largest number of cortical nerve cells is found. It contains more than a third of the entire cortex and also the largest volume of white matter. In contrast to other primates, the number of neurons in specific human frontal cortex areas is particularly striking, e.g. in area number 10. The anatomist

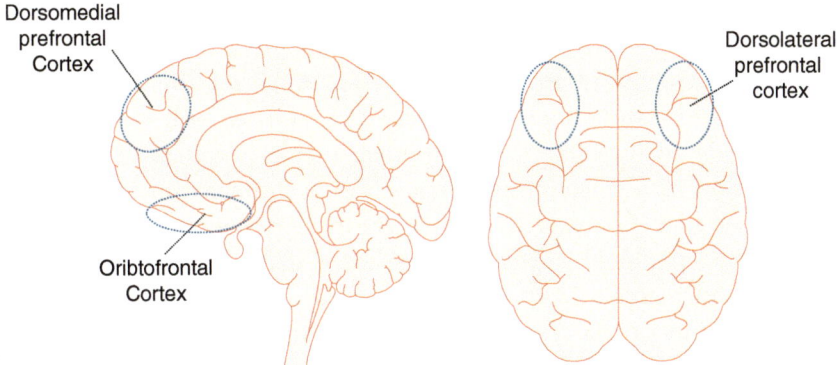

Dorsomedial prefrontal Cortex

Dorsolateral prefrontal cortex

Oribtofrontal Cortex

**Fig. 1.6** View of the brain cut along the midline (left) and from above (right). The areas of the neocortex relevant to specific human abilities are outlined. The working memory plays an important role in action planning and decisionmaking and is located in the dorsolateral (top lateral) region of the prefrontal cortex. Diverse neural networks that determine social behavior (empathy, altruism) and visceromotor functions pass through the dorsomedial cortex. Specific character traits that connect emotions emerging from the limbic system with the rational-cognitive functions of the frontal lobe are encoded in the orbitofrontal cortex. Lesions here result in apathy, but also disinhibitions (punning, indelicate joking) are observed

Korbinian Brodmann (1868–1918) divided the cortex into different fields based on histological differences more than 100 years ago. This classification is still used by doctors and researchers today. The most anterior prefrontal cortex and the cortical areas located below in the orbital frontal lobe (above the eye sockets) are of central importance (Fig. 1.6). For example, our working memory is located in the dorsolateral prefrontal cortex. It is large in comparison to other mammals (but not unexpectedly large in relation to the enlargement of the brain as a whole). Since this area takes up almost twice as much space in humans as in our closest relative, the chimpanzee, it is reasonable to look for a neuroanatomical correlate of higher cognitive functions here.

## 1.12 The Prefrontal Cortex Encodes Human Specific Properties

From the prefrontal cortex reflexive behaviors can be suppressed (as seen from disinhibitions in orbital cortex syndromes). This enables people to postpone immediate needs and plan their behavior for the future. The ability to interrupt the fully automated behavior still occurring in most animals gives people time to look for alternative approaches to complex problems that may require a different solution than the one they would intuitively prefer. For example, in case of acute danger, the innate or early learned behavior can be interrupted before it is carried out. While all monkeys run away in the same direction when they are warned of an attacker by the scream of a single animal in the group, humans do not show this uniform behavior. Some of us will pause for a moment and think about whether there might be a better escape route than the one specified by a certain person high up in the hierarchy when we are warned by a sign of danger. This function is typically associated with the prefrontal cortex. We must realize that we all genetically descend from these more intelligent people who, using their prefrontal neurons, made the wiser decisions in critical situations.

All primates lose the ability to think unconventionally when damage occurs to their prefrontal regions. For example, monkeys with injuries in this area cannot retrieve bananas that are visible behind a transparent wall, but they begin to push against the wall with their hand. Healthy monkeys and humans actually do this "thinking around the corner" without any problems. Unconventional behavior is therefore characterized not by the choice of the simplest, but by choosing the most promising path in the long term. It is a phylogenetically young property that has been observed not only in primates and elephants, but also in some birds. For example, crows come up with the idea of forming a hook from a straight wire with which food pellets can be removed from a narrow tube (a truly unconventional way of getting food). This complexity requires a certain minimum number of neurons. In fact, the bird brain contains about twice as many nerve cells per unit weight compared to primates, which are also packed at much higher density.

So the prefrontal cortex enables the famous outside-the-box thinking. In contrast to most animal species, we are particularly good at waiting, thinking and postponing acute needs such as hunger and thirst in order to think of alternative behavioral strategies that offer long-term advantages. Via interaction with other cortex areas and deeper brain structures, the prefrontal cortex enables us with a new form of biological intelligence, which is defined by collecting space-time information about various objects and interlinking them in our brain. This allows us to forge plans, the potential results of which we can reflect in our mind before they are ultimately set in action. Furthermore, some of the prefrontal cortex areas also permit us to act considerately and empathically. This creates socially adequate and emotionally intelligent behavior that benefits the entire group. However, these newly acquired abilities in evolution also allow humans to influence their environment in a way that calls the entire human existence into question (for example, by causing climate catastrophes or nuclear weapon explosions).

Taken together, the rule "the more the better" applies to neuron numbers in the prefrontal cortex, because the human brain makes us the superior species on this planet, at least if one evaluates superiority according to criteria such as: Who can develop complex sociological structures or political systems? Who survives at −50 ° and +50 ° outside temperature? Who builds an 800 m high house with 160 usable floors? (The Burj Khalifa in Dubai is actually that big).

## 1.13 Brain Performance in Comparison

In this chapter, I tried to show that the skills acquired over many millennia and passed down from generation to generation are coupled to a wide cerebral cortex, which in humans contains about 16 billion multiple layered and interconnected neurons. In this context, it should be clarified that not only 10% of all neurons are needed in our brain, as one can read again and again in popular scientific literature. On the contrary, the nerve cells present after development is finished are practically all integrated into functionally relevant neuronal networks. By the way, it would make no sense at all to unnecessarily supply 90% of unused brain substance with energy.

Are there differences between the sexes? Interestingly, the male neocortex is slightly larger and contains more neurons than the female. But it must be said that in today's living humans, brain size or cortex volume does not correlate well with general cognitive abilities. Even Albert Einstein's brain showed no special abnormalities in this respect. We assume that at normal variations in the number of neurons, the genetically determined and learning-modified number and strength of synaptic connections (i.e., the neuronal networks) make the decisive difference between human brains and not the overall number of neurons.

However, the increased connectivity and complexity of neuronal networks does not seem to be the only criterion for achieving high cognitive performance. In contrast, during the course of phylogeny, we have even experienced disconnection and specialization in some brain areas. For example, the region of the brain innervating the visual cortex in primates, the lateral geniculate body of the thalamus, projects to various other targets in cats and other mammals. The same applies to the thalamo-cortical projections of our general surface sensations, the somatosensitivity. Hence, in the course of phylogenetic brain development, there has been a reduction in connectivity in some areas, so that parallel processing of signals has been replaced by serial data processing.

In other brain regions, however, the number of neuronal connections has increased significantly. This is particularly visible in the neocortex, which has been able to connect strongly with subcortical nuclei. This is known as Deacon's rule (*large equals well connected*). In primates, the numerous direct connections of the motor cortex to the neurons in the spinal cord responsible for hand and finger movements are particularly noteworthy. These are much more extensive than in other species. In addition, during evolution, the control of brainstem nuclei by cortical neurons has increased significantly. This mainly affects our speech organs, that is, the fine control of nerve cells that innervate facial and masticatory muscles, as well as our tongue and laryngeal muscles.

The development towards individually controllable, serially connected neuronal networks makes many systems particularly vulnerable. The destruction of only one node in the network can lead to the failure of the entire functionality. Indeed, after cortex lesions in rodents, the resulting defects are significantly less pronounced than in humans or monkeys. This may also be due to the increased functional specialization of the two hemispheres. Human brains are more asymmetrically designed and the two hemispheres less connected when compared to other vertebrates. The high specialization of our cortex, therefore, allows the development of more complex and better controllable motor programs with higher sensitivity. This entails an increased risk of failure, however, as we will see in the following chapter.

---

### In a Nutshell

- It was only 50,000 to 80,000 years ago that man developed a complex symbolic language, a prerequisite for higher cognitive performance.
- Our working memory is located in the dorsolateral prefrontal cortex. This region plays an important role in our action planning and decision making.
- Cortical modules located in the dorsomedial cortex mainly influence social behavior (empathy, altruism) and vegetative-autonomic functions.
- The connection of emotions with rational-cognitive functions takes place in the orbitofrontal cortex.
- The most important function of the prefrontal cortex is to suppress automatic (reflexive) behavior. This enables us to postpone acute needs and make alternative plans for later execution.

# Further Reading

Briscoe SD, Ragsdale CW (2018) Homology, neocortex, and the evolution of developmental mechanisms. Science 362:190–193

Cadwell CR, Bhaduri A, Mostajo-Radji MA, Keefe MG, Nowakowski TJ (2019) Development and arealization of the cerebral cortex. Neuron 103:980–1004

Charvet CJ, Striedter GF, Finlay BL (2011) Evo-devo and brain scaling: candidate developmental mechanisms for variation and constancy in vertebrate brain evolution. Brain Behav Evol 78:248–257

Esteves M, Lopes SS, Almeida A, Sousa N, Leite-Almeida H (2020) Unmasking the relevance of hemispheric asymmetries—break on through (to the other side). Prog Neurobiol 192:101823

García-Moreno F, Molnár Z (2020) Variations of telencephalic development that paved the way for neocortical evolution. Prog Neurobiol 194:101865

González-Forero M, Gardner A (2018) Inference of ecological and social drivers of human brain-size evolution. Nature 557:554–557

Hagoort P (2019) The neurobiology of language beyond single-word processing. Science 366:55–58

Herculano-Houzel S (2020) Birds do have a brain cortex—and think. Science 369:1567–1568

Jarvis ED (2019) Evolution of vocal learning and spoken language. Science 366:50–54

Krubitzer L, Dooley JC (2013) Cortical plasticity within and across lifetimes: how can development inform us about phenotypic transformations? Front Hum Neurosci 7:620

Manger PR (2006) An examination of cetacean brain structure with a novel hypothesis correlating thermogenesis to the evolution of a big brain. Biol Rev 81:293–338

McGowan LD, Alaama RA, Freise AC, Huang JC, Charvet CJ, Striedter GF (2012) Expansion, folding, and abnormal lamination of the chick optic tectum after intraventricular injections of FGF2. Proc Natl Acad Sci 109:10640–10646

Moreno-Jiménez EP, Flor-García M, Terreros-Roncal J, Rábano A et al (2019) Adult hippocampal neurogenesis is abundant in neurologically healthy subjects and drops sharply in patients with Alzheimer's disease. Nat Med 25:554–560

Němec P, Osten P (2020) The evolution of brain structure captured in stereotyped cell count and cell type distributions. Curr Opin Neurobiol 60:176–183

Pennisi E (2019) How life blossomed after the dinosaurs died. Science 366:409–409

Reardon PK, Seidlitz J, Vandekar S, Liu S et al (2018) Normative brain size variation and brain shape diversity in humans. Science 360:1222–1227

Scott SK (2019) From speech and talkers to the social world: the neural processing of human spoken language. Science 366:58–62

Snyder JS (2019) Recalibrating the relevance of adult neurogenesis. Trends Neurosci 42:164–178

Stacho M, Herold C, Rook N, Wagner H, Axer M, Amunts K, Güntürkün O (2020) A cortex-like canonical circuit in the avian forebrain. Science 369:eabc5534

Striedter GF, Srinivasan S, Monuki ES (2015) Cortical folding: when, where, how, and why? Annu Rev Neurosci 38:291–307

Tosches MA, Yamawaki TM, Naumann RK, Jacobi AA, Tushev G, Laurent G (2018) Evolution of pallium, hippocampus, and cortical cell types revealed by single-cell transcriptomics in reptiles. Science 360:881–888

# 2

# Aging and Neurodegenerative Diseases: Why do Nerve Cells Die?

In the first chapter, I tried to explain why we need many neurons, particularly in the neocortex. Here, various new areas have been formed over the millions of years, which do not exist in other species. Billions of new neuronal connections have been created as a result of the increased number of nerve cells and cellular processes allowing an ever higher level of mental performance.

The neurons located in the frontal lobe, in particular, play a key role in those neuronal networks that generate human specific abilities, such as speech and extensive non-verbal communication through targeted innervation of facial muscles. Furthermore, impulse control and predictive planning distinguish us from most animals. The loss of cells necessary for these tasks in old age or in neurodegenerative diseases leads to corresponding failures, which sometimes make it impossible for us to survive in this demanding world. We lose our independence in everyday life and require external care. Although most neurons in the cortex usually perform their tasks for a long time, they succumb too early in Alzheimer's disease, and Parkinson's patients wish for nothing more than the recovery of neurons in the midbrain of their brainstem, so that they can control their facial and body muscles as usual.

The first part of this chapter will investigate the question of why and how nerve cells die at all. In addition to the natural aging process, the two most important neurodegenerative diseases, Parkinson's and Alzheimer's disease, are in the foreground. Both have an enormous social relevance: In Germany alone, the number of affected patients will increase to over three million (4 % of the population) by the middle of the century.

L. P. Klimaschewski, *Parkinson's and Alzheimer's Today*, https://doi.org/10.1007/978-3-662-66369-1_2

## 2.1  The Normal Aging Process

We are getting older. The average life expectancy was only 50 years a century ago, but today it is already over 90 years in developed countries. The reasons for this are first of all the successful push back of epidemics through vaccinations as well as improved hygiene and nutrition. The discovery of antibiotics has also contributed significantly to the effective fight against infectious diseases. A wide range of supplies ensures this ongoing development in many countries of the world. The demographic analysts therefore calculate a doubling of the number of people over 65 by 2030 in most industrial countries. In 2050, according to the World Health Organization (WHO), there will be around 2 billion people over 60 years old on Earth, many of them over 85 years old, making them the population group with the highest growth rate.

It must be assumed, therefore, that there will be a significant increase in age-related physical and mental illnesses. This includes memory disorders such as dementia, as well as degenerative joint and bone diseases. But why do we age at all? An explanation can be found in the evolution of man, because only a few people in the past lived beyond 80 years. It is important for the understanding of aging in general that after many years of life, most mammals can no longer provide their offspring with a decisive advantage in terms of their reproduction. Once the grandchildren are over 25 years old, evolution no longer matters. There is indeed a "grandmother effect", but unfortunately no "great-grandmother effect" (although great-grandparents can contribute to the dissemination of knowledge). In the course of human evolution, simply no genes have been selected that guarantee longevity beyond 70–80 years. The very old organism can actually become a competitor of the younger ones in the daily struggle for food and living space. So over the course of millions of years, no positive selection pressure built up on the very old mammal, as we saw in the first chapter in relation to younger people and brain development. The opposite is the case. There is a negative selection pressure due to the scarcity of resources.

In principle, under optimal genetic and environmental conditions, our brain can still function well after 100 years. But after 120–130 years there is no way to stop the natural decline. The oldest person so far, the Frenchwoman Jeanne Louise Calment, died in 1997. She was 122 years and 164 days old. In addition to the restrictions on mobility and memory, older people often have deficits in the sensory and cognitive functions. Concentration and attention are impaired. These *executive functions* are slowed down and no longer fully available. So their ability to quickly adapt to changing environments is limited. There are enormous differences between people of the same age group, however. While the "fluid" intelligence is reduced, the "crystallized" intelligence, that is our cognitive functions shaped by knowledge and experience, can be very well preserved in old age. This fact makes us optimistic because there must be ways to maintain mental and cognitive freshness even in old age.

Apparently, however, a special genetic predisposition is necessary for this. If the genetically determined conditions are favorable, it is possible to delay the beginning of neuronal degeneration by many years. This includes, in addition to the observance of the classical lifestyle factors (movement, balanced nutrition, little alcohol, no nicotine), a high level of mental activity. I will go into this in more detail in the third chapter. It is striking that the training effect of intellectual efforts is comparable to physical workouts. It even lasts longer, particularly, when we were mentally active in our youth.

To make the brain as fit as possible for old age, one should therefore regularly exercise the mind and intensively exchange stimulating content with other people, starting in childhood and adolescence. Today we know that the effects of this early, continuous mental work on the structure and function of our neuronal networks are strong and cannot be achieved by any substances supplied from the outside, for example, by the popular memory enhancers. We have to work on our cognitive reserve that can be maintained and expanded in old age by practicing creative skills, e.g. learning a new language or a musical instrument.

**In a Nutshell**

- In 2050, 2 billion people on Earth will be over 60 years, many of them over 85 years old.
- Nature has given us a genetic basic equipment that is programmed for a maximum life expectancy of 120–130 years.
- Our intelligence, shaped by knowledge and experience, can be fully preserved in old age.
- In addition to physical exercise and balanced nutrition, a high level of mental activity is required, which should begin as early as possible in life.

## 2.1.1 Mechanisms of Cellular Aging

Like many other cells, neurons are lost during aging. Already after the 20th birthday we lose daily at least 1000 of them (other estimates expect a loss of up to 100,000 neurons per day). Many neurons do not die but shrink with age, i.e. their cell body's volume is reduced. This leads to a restriction of cellular functions, which is, however, reversible in contrast to cell death and called senescence. The molecular and genetic mechanisms underlying cellular senescence are currently being intensively investigated. The results available to date show that cell aging is primarily due to an accumulation of irreparable mutations in mitochondrial and nuclear DNA. Neurons as post-mitotic cells are particularly sensitive to these changes because they can no longer divide and their DNA can therefore no longer be replicated like in dividing cells.

**DNA Telomeres Determine the Number of Cell Divisions**

Another important cause of aging is the shortening of telomeres. These are the DNA sections at the end of the strands of DNA (chromosomes) that do not contain any genetic information, but protect the DNA from degradation (Fig. 2.1). They become shorter with each cell division. Just like a candle burning down, the length of telomeres in dividing cells therefore decreases steadily. This also affects the glia in our brain, but not the post-mitotic neurons. The length of telomeres is measured in base pairs (i.e. in DNA units). Newborns have a telomere length of around 10,000 base pairs, someone over 40 already has about a third less, and over 60 years there are only about 5000 base pairs left. Since a cell can no longer divide when it reaches a critical telomere length, but instead enters programmed cell death (discussed below), nature has installed an intrinsic mechanism for aging in all dividing cells using telomeres. However, there is also the possibility of repair: An enzyme that counteracts telomere shortening has been found particularly in epithelial and lymphatic cells. This so-called telomerase is active in stem cells, and plays an important role in tumor growth.

A low activity of telomerase promotes cellular aging and eventually, together with other factors, terminates further cell division (proliferation). This leads to cellular senescence. In addition, the anti-oxidative effects of telomerase are of interest. They have nothing to do with the extension of chromosomal endings, but inhibit the formation of oxygen radicals in mitochondria as shown by treatment with telomerase activators. In addition, the latter reduce the levels of pathological proteins and improve the

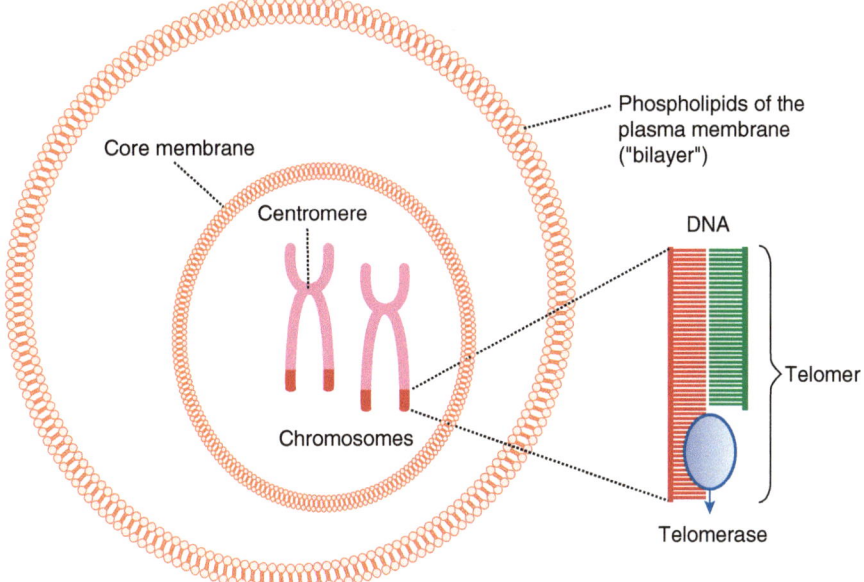

**Fig. 2.1** Schematic representation of a cell. Both are surrounded by a phospholipid bilayer membrane. The 46 chromosomes (DNA strands) are located in the nucleus. They each consist of two arms (chromatids) held together at the centromere and contain the genetic material (DNA). Telomeres form the ends of the chromosomes. They resemble the sealed ends of a shoelace and shorten with each cell division, but can be repaired by an enzyme called telomerase

symptomatology in animal models of Parkinson's disease. Attempts have also been made to slow down aging by increasing telomerase activity. This should stop the typical aging phenomena such as wrinkled skin, white hair or reduced vision.

## Aging Cells Maintain a Chronic Inflammation

In addition, aging research is working on methods to remove senescent cells, as they can cause inflammation over time and damage the tissue as a result. These so-called "senolytics" aim at specific anti-apoptotic proteins that are highly expressed in aging cells and drive these cells into programmed cell death (apoptosis). By pharmacological targeting of the anti-apoptotic molecules the senescent cells can be eliminated.

Another possible target is kidney-associated glutaminase. This enzyme increases the intracellular ammonium concentration through glutamine cleavage and thus normalizes the pH value of affected cells. A permanent drop in the pH induced by glutaminase inhibitors causes senescent cells to die. However, it should be noted that senescent cells may also have useful biological functions. They are involved in tissue repair and their pharmacological destruction could cause serious side effects.

In addition, neurons and muscle cells, the myocytes, are subject to processes of aging, which are less relevant in dividing cells. This includes an accumulation of protein complexes that can no longer be removed from the cell. In mitotically active cells, protein aggregates are distributed to the daughter cells formed after cell division, i.e. they are diluted over time. In contrast, such deposits are enriched in neurons and other postmitotic cells. In the past, it was thought that protein deposits themselves trigger a neurodegenerative disease. Today, many researchers believe the opposite, i.e. that they probably act to delay the disease, as the cell tries to form aggregates to get rid of the toxic, still largely soluble precursors of those proteins forming aggregates. This will be discussed in more detail in the paragraphs on Parkinson's and Alzheimer's disease later.

Some proteins that occur more frequently in neurodegenerative diseases and are cell-damaging have already been well studied, for example, α-synuclein in Parkinson's or tau protein in Alzheimer's disease. It is conceivable that in the future more substances will be discovered that may play a pathogenetic role in these disorders. Apparently, neurons over a certain age can no longer regulate the degradation of proteins, that are continuously produced at a high rate.

### In a Nutshell

- Cell aging is primarily due to an accumulation of irreversible mutations in the genetic material, deoxyribonucleic acid (DNA), in the cell nucleus and in mitochondria.
- Neurons are particularly sensitive to mutations because, unlike dividing cells, they cannot replicate their DNA and their ability to repair genes is limited.
- Nerve cells also age through the accumulation of various proteins that are continually produced in large amounts but can no longer be efficiently eliminated.
- Senescent glial cells maintain a chronic inflammation in the brain.

## The Importance of Protein Homeostasis for Cellular Aging

All proteins present in a cell at a given time form the proteome. A single cell uses approximately half of the 20,000 genes present in the nucleus to encode proteins, i.e. about 10,000 different proteins are produced in each cell. Estimates suggest that a cell contains a total of about $10^{11}$ proteins. This means that there are more proteins in a human being than stars in the Milky Way. It is therefore a Herculean task to find the changes in the proteome that are responsible for senescence and neurodegenerative diseases. After the Human Genome Project was completed a few years ago, this is certainly one of the biggest challenges for medical research with regard to aging.

The protein homeostasis, also called proteostasis, describes the synthesis, folding, distribution and degradation of proteins necessary for all cell functions (Fig. 2.2). So the entire journey of a protein from its production at the ribosome, the transport within the cell to its final location until its breakdown in lysosomes or in proteasomes, the cellular recycle bins, is of utmost importance. In addition, it is also highly relevant that a protein is correctly folded after synthesis, i.e. its spatial structure must fit so that it can interact with other molecules in the cell in a meaningful way. The folding of a long chain of amino acids usually involves other proteins. These are referred to as chaperones and play an important role as "nannies" for other proteins, as around one third of all proteins in the cell have the tendency to not fold correctly.

Incorrectly configured proteins can be excreted or they are recognized by the endoplasmic reticulum (ER) and degraded in the proteasome. Special ER enzymes are required for this task, for example, the membralin protein. They together form the ERAD complex, an abbreviation for **ER-A**ssociated **D**egradation. If the ER is overloaded with incorrectly folded proteins in aging postmitotic cells, the Unfolded Protein Response (UPR) sets in. This cellular reaction tries to restore the normal state of the ER and thus protects the cell.

If the overload of the ER (ER stress) persists over a longer period of time, proteins localized in the membrane of the ER trigger specific signaling cascades that shut down protein synthesis at the ribosomes and lead to DNA strand breaks in the cellular nucleus. As a result, the axonal myelin sheaths are lost (demyelination) and the axons lose their continuity (fragmentation). In the end, the cell dies by apoptosis (see Sect. 2.1.2). A long-lasting overload with tau or α-synuclein proteins, which typically occur in neurodegenerative diseases, could trigger ER stress.

## Self-cleaning of Nerve Cells

In addition to the limited functions of enzymes that are involved in the aforementioned UPR (unfolded protein response) or UPS (ubiquitin-proteasome

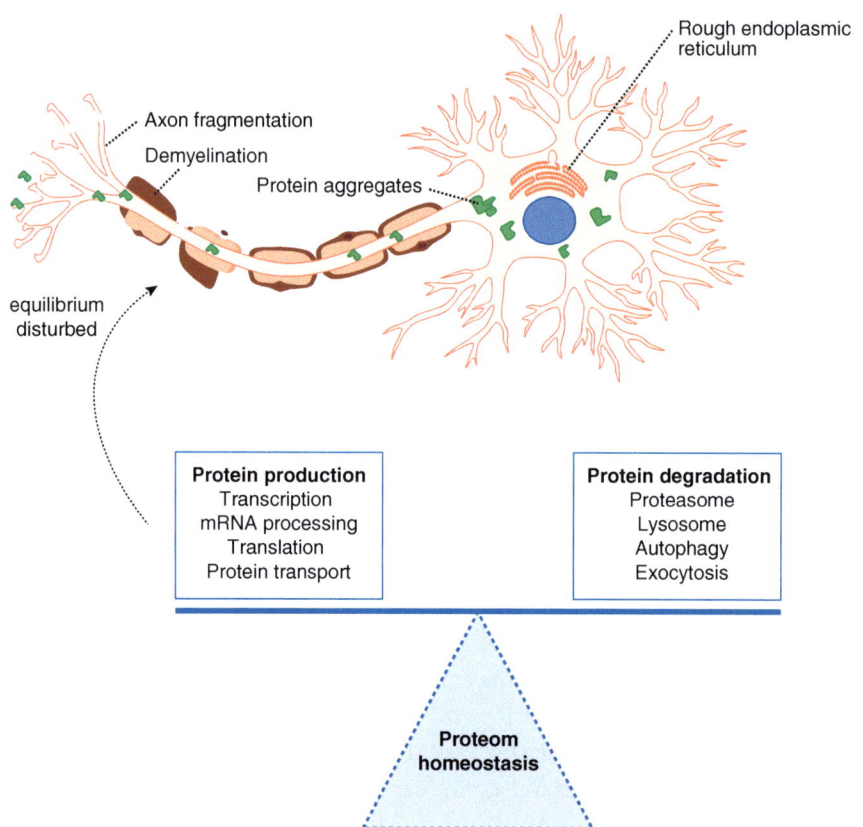

**Fig. 2.2** The balance of protein synthesis and degradation in cells is controlled by various processes at all subcellular locations (nucleus, cytoplasm and cellular extensions). In the context of neurodegenerative diseases, problems with protein degradation arise regularly, e.g. defects in the ubiquitin-proteasome-pathway or in the lysosomal system can be observed. Moreover, the removal of whole organelles (autophagy) may be disturbed. This leads to protein fibrils and aggregates, which weaken the lysosomal system or cause cellular stress, for example in the endoplasmic reticulum (the so-called ER stress)

system), "self-eating" (autophagy) plays a decisive role in disturbances of protein homeostasis. In this process, which is very important for normal cell metabolism, protein complexes and even entire organelles are removed by intracellular degradation or by secretion, known as exocytosis. In neurons the demands on this system are particularly high due to their usually long cellular extensions. Axons can be over one meter in length and metabolically very active, especially when they have to generate up to 50 impulses per second.

In macroautophagy, the structures to be disposed of (protein aggregates, pathogenic agents, mitochondria, peroxisomes) are enclosed by a newly formed double lipid membrane (Fig. 2.3). This creates a large transport vesicle, the autophagosome, which in neurons retrogradely reaches the cell body from dendrites and axons. There it fuses with one of the numerous lysosomes, whose acidic peptidases break down the content of autophagosomes and recycle the resulting amino acids, thus contributing to protein

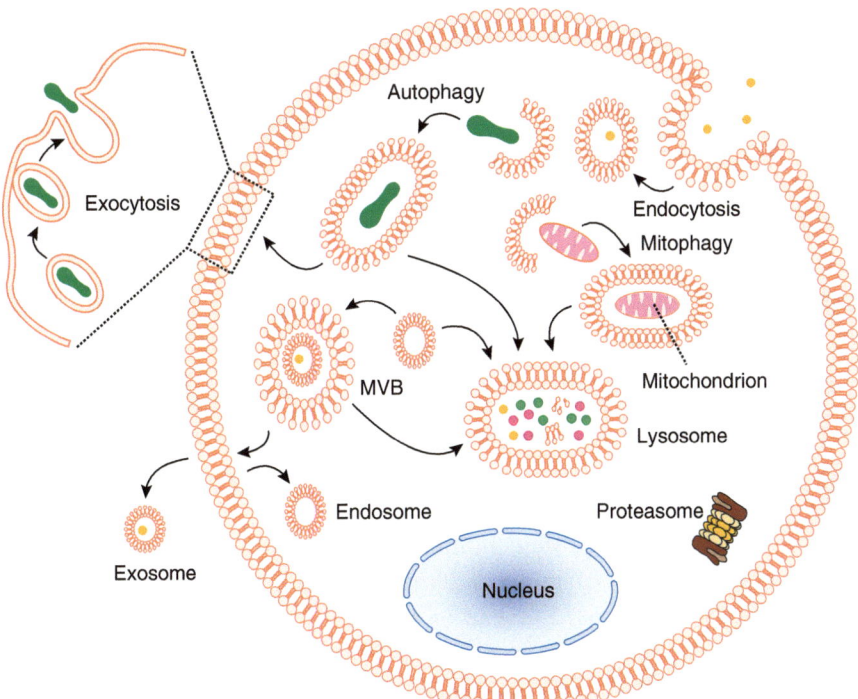

**Fig. 2.3**  A cell can take up various substances (yellow dots) from the outside by endocytosis (after invagination of the plasma membrane), pack them into vesicles (endosomes), and distribute them within the cell. In addition, protein aggregates (green), but also whole organelles (such as mitochondria, pink) are enclosed by intracellular membranes as needed. These processes are referred to as autophagy or mitophagy. Autophagosomes in turn fuse with lysosomes, which break down their contents using protein-, sugar- and fat-splitting enzymes. This makes individual amino acids, glucose or lipids available for the cellular metabolism again (most cytoplasmic proteins, however, are degraded in the proteasome). In addition, smaller vesicles are detached from the endosomal membrane towards the inside (as in endocytosis). Several of those form the multi-vesicular bodies (MVBs). Various substances are also released directly into the extracellular space by exocytosis or secreted in the form of another type of small vesicles (exosomes, structures not drawn to scale)

biosynthesis. If the cell has only few nutrients left, for example in a state of hunger, this mechanism is of crucial importance in order to regenerate the necessary amino acids by breaking down muscle proteins. This process is called induced autophagy.

It is also possible that lysosomes take up cytoplasmic substances or organelles directly via membrane invaginations as part of micro-autophagy. Transporters located in the lysosomal membrane my help in this process, which is then called chaperone-mediated autophagy. The lysosomes are essential for the degradation and recycling of extra- and intracellular substances. They contain around 60 hydrolases, including those that can cleave sugar chains (polysaccharides), lipids or nucleic acids. Mutations in genes necessary for autophagy in neurons can lead to axon degeneration or even cell death. Proteins and organelles accumulate, bind to each other and aggregate. The 76-amino acid protein ubiquitin, which is used as a marker for the planned disposal of a protein in the proteasome, is often found in these aggregates.

The coupling of ubiquitin to proteins is a complicated process controlled by several enzymes of the above mentioned UPS, which in mutated form may also trigger a neurodegenerative disease. I will come back to this later in this chapter. Although it can be considered certain that most neurodegenerative diseases must be classified as multifactorial, the occurrence of protein aggregates is often defining for a specific pathology and attributable to problems in protein degradation or autophagy. The pharmacological stimulation of this process may have a therapeutic effect.

Maintaining cellular protein homeostasis is, therefore, a mammoth task that requires a variety of molecular interactions, particularly in the endosomal system of neurons. Because of the elementary importance of transport and degradation processes for the maintenance of cellular survival, even minor disruptions can lead to significant restrictions of cellular functions which may result in cell death. In genetic screens that are supposed to detect disease-causing (e.g. mutated) DNA sequences, changes are frequently detected that are involved in protein biosynthesis, folding, intracellular transport or protein degradation. Interestingly, a number of lysosomal storage diseases are also characterized by pronounced neurodegeneration, and vice versa, neurodegenerative diseases show lysosomal dysfunctions. These interactions underscore the importance of protein homeostasis for neuronal survival in aging and in disorders leading to neuronal degeneration.

**In a Nutshell**

- A single cell produces about 10,000 different proteins.
- Intracellular protein homeostasis is of great importance. It includes synthesis, folding, distribution and degradation of proteins.
- Chaperones are proteins that serve as scaffolds for the correct folding of other proteins.
- Incorrectly configured proteins are recognized by the endoplasmic reticulum (ER) and degraded in the proteasome. If the ER is overloaded, protein synthesis is shut down at the ribosomes and the cell dies (ER stress).
- In addition to the ubiquitin-proteasome system (UPS), the lysosome is essential for the degradation and recycling of extra- and intracellular substances.
- Autophagy allows the removal of aging organelles and larger molecular complexes. It plays an important role in neuronal metabolism, especially in aging.
- The occurrence of protein aggregates defines neurodegenerative diseases, which can often be traced back to defects in protein transport and/or protein degradation.

## Two Sides of a Coin: Oxygen Radicals

For a long time oxidative processes have been at the center of aging research. It was assumed that oxygen radicals are the decisive drivers of cellular senescence. Until a few years ago, it was popular to take anti-oxidant vitamins C and E as pills in order to intercept radicals and thus stop cell aging (unfortunately without measurable success). Oxidative stress cannot be avoided in aerobic life forms, because reactive oxygen species (ROS) and also reactive nitrogen species (RNS) are continuously produced in the course of metabolic processes, particularly at mitochondrial and endosomal membranes (Fig. 2.4 and 2.10).

Reactive nitro compounds, for example the peroxynitrite anion ($ONOO^-$), are formed by the inducible nitric oxide synthase (iNOS2). The gas NO (nitric oxide, that is also released as a neuromodulator in the brain) quickly diffuses into the environment and binds to the enzyme guanylate cyclase in target cells, which then forms cGMP as a signal molecule (*second messenger*). In addition, astrocytes and microglia are activated and inflammatory reactions started in the brain. Nitro compounds modify the amino acids cysteine and tyrosine and thus change the function of proteins. However, it is important to emphasize that the production of oxidants is essential for many cell types, for example for macrophages and neutrophilic granulocytes, which are supposed to destroy bacteria or tumor cells by oxidative processes as part of the innate immune response.

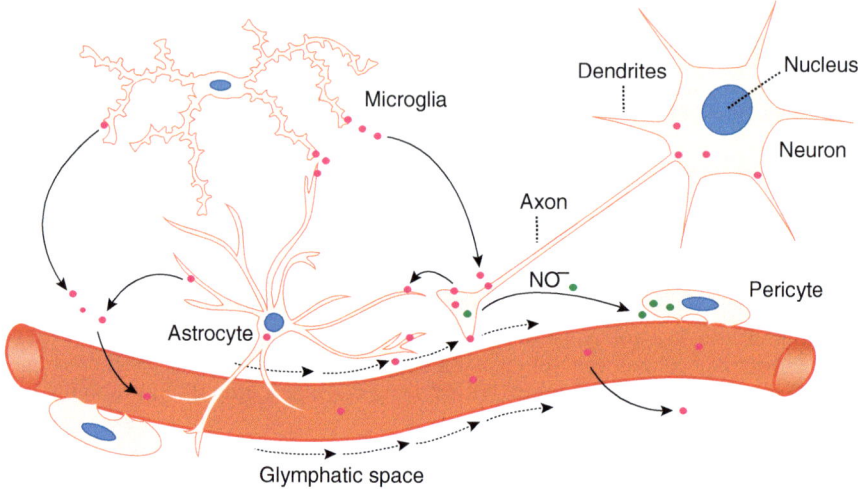

**Fig. 2.4** Neurons and astrocytes are in close contact with blood vessels and together form the neurovascular unit. Astrocytic processes are closely packed around capillaries (only some glial processes are indicated here for the sake of clarity). The perivascular area bounded by astrocytes is called the glymphatic space (in analogy to the glia and the lymph in the body; however, there is no lymph in the brain), in which a lively exchange of substances between blood vessels, astrocytes and neurons can be observed (red dots). Microglial cells, which are comparable to the monocytes in the blood, release inflammatory mediators (cytokines), particularly in the elderly, under pathological conditions. In addition, the pericytes sitting on the vessel wall play an important role because nitric oxide ($NO^-$, green dots) released from synaptic contacts leads to the dilation of blood vessels in the vicinity of activated nerve cells via its relaxing effect on these cells.

ROS are highly reactive chemicals, among them the superoxide anion radical ($O_2^-$), hydrogen peroxide ($H_2O_2$) and the free hydroxyl radical (OH). The latter can link fatty acids in the cell membrane together, so that the lipid double layer is destroyed. In addition, ROS endanger our DNA because they cause direct base damage. Furthermore, highly reactive aldehydes (e.g. the 4-hydroxynonenal, HNE) bind to proteins and interfere with their correct folding. Superoxide radicals and hydrogen peroxide also lead to the cross-linking of proteins, which lose their physiological function as a result. This can be explained by the fact that the amino acids normally located in the water-repellent (hydrophobic) core of a protein are brought to the surface by oxidation and bind to hydrophobic sections of adjacent proteins.

In this way, protein aggregates are formed, which are deposited in the cytosol and referred to as lipofuscin or as age pigment. In addition to protein oxidation, other protein modifications can also lead to aggregation, e.g. the reaction of proteins with sugar (glycoxidation). This coupling of monosaccharides to free amino groups of various proteins plays an important role in the formation of aggregates in the context of Parkinson's and Alzheimer's disease. However, it is still largely unclear how and to what extent these aggregates make nerve cells age or die, and whether the modifications are possibly reversible.

## Endogenous Radical Scavengers Protect Our Nerve Cells

Highly reactive oxygen compounds that can be formed by metabolic enzymes such as xanthine oxidase and enzymes of catecholamine degradation must be detoxified by cellular defence systems, and damage induced by oxidants is constantly repaired. Low-molecular anti-oxidants (radical scavengers) are, for example, uric acid, vitamin C (ascorbic acid), vitamin E (tocopherol) and beta-carotene (provitamin A). The enzymes superoxide dismutase, catalase and glutathione peroxidase play important roles as endogenous antioxidants. Plant substances such as carotenoids and polyphenols (resveratrol, curcumin, or flavonoids) also have an anti-oxidative effect, inhibit inflammatory reactions and stimulate the release of neurotrophic factors that promote neuronal survival and process outgrowth.

In the CNS, it is primarily the astrocytes that protect nerve cells from oxygen radicals by releasing glutathione and providing the glutathione precursor (CysGly). In the context of neurodegenerative diseases, this important task of the astroglia can often no longer be carried out. In addition, the microglial cells phagocytose less effectively and change into a senescence-like state from which it can no longer renew itself.

The modified cells are then referred to as "A1 astrocytes" and "M1 microglia", respectively. The prevention of glia conversion significantly reduces neuronal degeneration in several animal models. Changeable, "plastic" glial cells can also be found in the younger brain and in a variety of lesions following the release of toxic substances or inflammatory mediators (TNFα or IL-1β). However, these cells then remain in the location of the disease and do not spread, as in old age or in neurodegeneration, over large parts of the brain.

**In a Nutshell**

- The area delimited by astrocytes around the small vessels is called the perivascular space. Various substances are exchanged between capillaries, astrocytes and neurons.
- The microglia in the brain, comparable to the monocytes in the blood, produces inflammatory mediators in old age and in neurodegenerative disorders, among them several interleukins and the tumor necrosis factor, TNF-$\alpha$.
- In addition to cytokines reactive oxygen compounds (ROS) are constantly formed at mitochondrial and endosomal membranes.
- Oxidative stress damages DNA, lipids and proteins. However, the production of ROS cannot be completely avoided in aerobic life forms.
- In addition to endogenous anti-oxidants (e.g. vitamin C and E, glutathione, superoxide dismutase, catalase), carotenoids and some polyphenols are protective for neurons.

## Chronic Inflammatory Processes in the Brain

Taken together, an imbalance of oxidative and anti-oxidative mechanisms over a longer period of time leads to chronic stress, which is harmful to postmitotic and proliferating cells alike. However, tissue and organ damage also occurs because aging cells release a number of inflammatory mediators (cytokines), which act on adjacent cells and can send them into senescence as well, referred to as the *bystander* effect. These factors include interleukins (in particular IL-1$\beta$, IL-6 and IL-8) as well as various other cytokines, e.g. tumor necrosis factor (TNF-$\alpha$). In an organ that is actually immunologically privileged like our brain, the mechanisms of the immune system are activated in the sense of a sterile inflammation in old age.

Immune cells are not only the source of inflammatory molecules but also produce enzymes involved in cellular aging. Particular attention has been paid in recent years to the metalloproteases (MMPs) released by inflammatory cells and microglia. MMPs are zinc- and calcium-dependent endopeptidases that cleave extracellular matrix molecules (laminin, collagen, fibronectin, etc.). They are mainly found in connective tissue, but also in the brain. A natural inhibitor of these proteases, TIMP-2, is thought to be responsible for the rejuvenation effect after blood plasma transfusion. In these experiments, which attracted much attention in 2017, the memory of old mice improved when they were injected with the blood plasma of young people. It is assumed that metalloproteases released from activated microglia are responsible for at least some aspects of the aging process, for example by loosening synaptic contacts that need to be stabilized by extracellular matrix proteins in the adult brain.

Furthermore, inflammation is driven by lifestyle, particularly in aging. Many of the older population in the western industrialized world suffer from the so-called metabolic syndrome. This combination of increased body mass index (BMI), type 2 diabetes, high blood pressure and disorders of lipid metabolism results in the release of several of the above mentioned inflammatory mediators by abdominal fat cells (adipocytes). The cytokines then cross the blood-brain barrier, promote the aging process and thus accelerate neurodegeneration.

However, it must be emphasized that inflammatory mediators also play a central role in acute tissue reactions, e.g. following injuries or in the control of tumor growth, and are indispensable in the context of wound healing. This Janus-faced characteristic of the immunological messenger substances is observed in most organs and, in particular, in the non-regenerative CNS. It is the duration and intensity of these inflammatory reactions that are decisive, as prolonged cytokine action can lead to considerable tissue changes. It is likely that some of the neurological symptoms of the up-coming "long-covid" crisis are due to such immunological mechanisms.

## 2.1.2 Neuronal Cell Death

After aging comes death. This applies to the entire organism as well as to its individual components, the cells, which the pathologist Rudolf Virchow already recognized in 1855 (*omnis cellula e cellula*). What are the specifics of neuronal cell death ? Unlike direct cell damage caused by external influences, genetically determined suicide programs play an important role in the aging brain and in neurodegenerative diseases.

In principle, programmed cell death, called apoptosis, is as important for tissue homeostasis as cellular proliferation, because in most organs (with the exception of the heart and brain) new cells are constantly being formed. These cells have to be removed permanently to prevent an enlargement of our organs. Neurons and heart muscle cells, on the other hand, are postmitotic, i.e. they survive the whole life, and are protected from the built-in cell death mechanisms.

However, neurons do die and they do this by more than ten different types of death, the molecular mechanisms of which partially overlap. Programmed cell death plays an important role in brain development to adjust the number of nerve cells to the number of possible target cells that they control. Hence, in the developing nervous system, the availability of neurotrophic factors taken up from the target cells ensures the survival of neurons innervating them.

This is a suitable mechanism to determine the required number of neurons in the brain, because neurons are produced in excess and those that cannot be integrated into functional neuronal networks will be removed. On the other hand, some neuronal networks are created several times to ensure the functionality of brain centers necessary for survival, for example, regulating the circulatory and musculoskeletal systems.

Therefore, a comprehensive back-up of nerve cells is present at birth, which can compensate for loss of neurons in the aging brain. In fact, more than 50% of dopamine-producing nerve cells in the midbrain will be affected in Parkinson's disease once the patient notices any motor symptoms.

## How do neurons die?

The prototype of a molecule that ensures neuronal survival is the nerve growth factor discovered by Rita Levi-Montalcini and colleagues in 1952. It prevents neuronal apoptosis. This term (Greek for "falling off") was coined by John Kerr in 1972 and describes the suicidal cell death during organ development when neurons have to be removed because they do not find appropriate target cells. However, some cell types, for example the small interneurons in the cortex, die independently of such trophic support. Instead, they use an intrinsic timer.

During apoptosis, the cell first shrinks, then the cellular nucleus disintegrates and the genetic material condenses. This leads to cleavage of the genetic material. DNA fragments can be made visible biochemically, and are histologically detectable by the TUNEL method. In addition, the cell characteristically forms so-called apoptotic bodies, which release a variety of molecules and indicate the final phase of cell death.

In contrast, after traumatic lesions neurons start to swell. The plasma membrane breaks open, cell organelles enter the environment and the cell dies. This process is referred to as necrosis (as opposed to apoptosis). Necrosis is accompanied by a local inflammation and by macrophages removing the cell debris. Based on current research analyzing cell death in neurodegenerative diseases, the transition between the two forms of cell death appears to be fluid. Interestingly, necrosis may also be controlled genetically and then referred to as "regulated necrosis". The neuronal cell death in neurodegenerative diseases is clearly characterized by aspects of both mechanisms, apoptotic and necrotic cell death.

The cellular changes in apoptosis usually occur as a sequence of activations of protein cleaving enzymes, the caspases. These are cysteine-aspartic proteases that can be initiated by cellular receptors on the plasma membrane or by endogenous, cell-internal mechanisms, which are located in the mitochondria.

As discussed above, mitochondria are the cell's power plants that produce energy carriers such as adenosine triphosphate (ATP) by means of oxidative phosphorylation. They are 0.5–1.5 $\mu$m large organelles that are enclosed by a double membrane and accomodate their own DNA. The mitochondrial membranes contain the electron chains necessary for cell respiration as well as the cytochrome P450 oxidases. Mitochondria multiply by growth and division, they fuse and are subject to a strict quality control. Their number and size are constantly adapted to the energy needs of the cell. For example, muscle cells, sensory neurons or oocytes contain a large number of mitochondria. Cytochrome C is released from them during apoptosis, which activates caspase-9 and thus starts the cell death program. In contrast, caspase-8 and caspase-10 are activated during apoptosis triggered exogenously. The pro-apoptotic members of the Bcl-2 protein family play an important role as well, since Bax can destroy the outer mitochondrial membrane by forming pores. Activated caspases or Bax may also lead to rupture of lysosomes. As a result, the released lysosomal enzymes (cathepsins, hydrolases, DNAses) lead to self-digestion (autolysis) of the cell.

The apoptotic process called pyroptosis is triggered by inflammatory mediators. In glial cells and possibly in neurons it is induced after activation of caspase-1 in the so-called inflammasome, a cytosolic multiprotein complex of the innate immune system. The protein gasdermin D is then formed, which leads to pores in the plasma membrane, resulting in calcium being able to flow unhindered into the cell and the cytoplasm being able to leave the cell, i.e. it runs out. Increased membrane permeability can also be triggered by other stimuli leading to cell death. In particular, in Parkinson's disease, the ferroptosis, an iron-dependent form of regulated apoptosis, plays a role as well (Table 2.1).

**Table 2.1** The comparison of important neuronal cell death forms based on their triggers, mediators and effects. In contrast to intrinsically initiated neuronal apoptosis, e.g. by a lack of neurotrophic factors (NTFs), extrinsic apoptosis is induced by tumor necrosis factor (TNF) (or by other members of the TNF family). Cytokines like TNF and activators of receptor-interacting protein kinase 3 (RIP3) initiate necrosis, i.e. cell death with pore formation in the plasma membrane. Lysosomal autolysis is initiated by lysosomal membrane permeabilization (LMP). However, DNA fragmentation into about 200 base pairs long DNA strands, the DNA ladder, only occurs during programmed cell death, which is accompanied by the complete clearance (phagocytosis) of the neuron by microglia and macrophages. ROS = reactive oxygen species, Ca = calcium, Fe = iron

|  | Initiators | Mediators | DNA Ladder | Pores | Effects |
|---|---|---|---|---|---|
| Intrinsic apoptosis | NTF Withdrawal | Caspase-9 | Yes | Mitochondria | Phagocytosis |
| Extrinsic apoptosis | TNFα, FasL | Caspase-8 | Yes | No | Phagocytosis |
| Regulated necrosis | TNFα | RIP3 | No | Plasma-membrane | Necrosis |
| Lysosomal autolysis | ROS, Ca$^{2+}$ | LMP, Cathepsin | No | Lysosomes | Necrosis |
| Pyroptosis | Cytokines | Caspase-1 Gasdermin | No | Plasma-membrane | Necrosis |
| Ferroptosis | Fe$^{2+}$ | Fe$^{2+}$, ROS | No | No | Necrosis |

The target proteins (substrates) of caspases are not all known, but it is assumed that up to 5% of all cellular proteins are cleaved by them. This includes the amyloid precursor protein (APP), which is important in Alzheimer's disease and has a suitable cleavage site for caspases. It is therapeutically relevant that cell death can be delayed in animal models of neurodegenerative diseases by caspase inhibitors. Furthermore, the antibiotic minocycline protects neurons from death because it inhibits the mitochondrial release of the aforementioned cytochrome C. As mentioned above, various neurotrophic factors, including neurotrophins (NGF, BDNF, NT-3), but also fibroblast growth factors (FGFs), inhibit the activation of caspases and thus prevent neuronal cell death. This is not only relevant during neuronal development, but also in neurological disease.

**In a Nutshell**

- There are more than ten different types of neuronal cell death. During development, in old age and in neurodegenerative diseases, apoptosis is often observed, i.e. the cell shrinks and the nucleus disintegrates.
- In necrosis, the neuron swells up and the plasma membrane breaks open. Macrophages then eliminate the cell debris.
- The cellular changes in apoptosis occur as a result of enzyme activation. These are primarily the caspases that can be activated by cellular receptors on the plasma membrane, but also from within the cell.
- Mitochondria release cytochrome C during apoptosis, which in turn activates caspase-9. In exogenously initiated apoptosis, caspase-8 and caspase-10 play the dominant roles.
- Cell death occurring in old age and in neurodegenerative diseases can be delayed by caspase inhibitors or by neurotrophic factor treatment.
- Pyroptosis leads to the release of cytokines and cell death of glial cells by activation of caspase-1 in the inflammasome, a multiprotein complex of the innate immune system.

## 2.1.3 Blood Supply of the Aging Brain

The involvement of blood vessels in the aging processes of the brain and in neuronal degeneration has been underestimated so far. It has been known for some time that circulatory problems in the CNS correlate with the development of memory disorders. A mild cognitive deficit is most often caused by insufficient blood flow to the brain.

The blood supply of the over 80 billion neurons is ensured by a network of blood vessels (arteries, capillaries and veins) of around 650 km in length. This means that no nerve cell is more than 15 micrometers ($\mu$m) away from a blood vessel (a human hair is approximately 50 $\mu$m thick). The capillary system of the brain is very pronounced, since it supplies one fifth of the total energy carriers and oxygen required by the organism, although the brain only accounts for one fiftieth of the body weight (see Chap. 1).

Another feature of the CNS is the blood-brain barrier. It prevents the uncontrolled passage of cells and, in particular, of water-soluble (polar) molecules across the inner layer of blood vessels, the endothelium (smaller, lipophilic substances can diffuse, however). Endothelial cells exhibit *tight junctions*, i.e. band-like connections of the plasma membranes of adjacent cells. With the exception of the capillaries, all vessels (arterioles, arteries, venules, veins) also have a layer of smooth muscle cells (myocytes) with some pericytes lying on the outside, which are of crucial importance for the development of the blood-brain barrier (see Fig. 2.4).

Between nerve cells and blood vessels are the extensions of the astrocytes mentioned above, which are also connected by tight, but channel-like contacts (*gap junctions*). Together, endothelium, astrocytes and neurons form the **neurovascular unit**. The endothelium acts as the actual barrier, since it prevents the passage of large macromolecules and whole cells, but also of pathogens (e.g. bacteria).

## Barrier Disorders are Not Uncommon in the Elderly

The perivascular area bounded by astrocytes, also known as the Virchow-Robin space around blood vessels, is of particular importance for the maintenance of brain homeostasis. This ensures that well-defined concentrations of ions and transmitters are present in the cerebrospinal fluid (CSF). The Virchow-Robin space contains the glial-lymphatic or glymphatic system, which is shown in Fig. 2.4. The term is somewhat misleading, as there are clear differences in the composition and function of the CSF compared to the body's lymph.

The glymphatic space expands up to 60% during sleep, resulting in a faster removal of unnecessary proteins, including β-amyloid in Alzheimer's disease. The system of Virchow-Robin spaces apparently plays a major role in the context of neurodegeneration, particularly at night, as the distance between the brain convolutions increases measurably when sleep is reduced by one hour, i.e. brain atrophy accelerates.

In addition to freely diffusible oxygen, our brain also requires a constant supply of sugar, which passes through pore-forming proteins (glucose transporters) from the serum into the neurons via the capillary endothelium. The nonspecific, vesicle-dependent transport via endocytosis (also called micropinocytosis) is missing in the endothelial cells of the brain, otherwise they would not be able to fulfill their barrier function. An interruption of the cerebral circulation, therefore, leads to unconsciousness within seconds and to permanent damage after a few minutes due to oxygen and sugar deficiency. This underlines the absolute dependence of the brain on glucose as the most important energy source.

There is increasing evidence that blood-brain barrier disruptions and reduced drainage of cerebrospinal fluid into the venous system play a key role not only in the general aging processes, but also in the development of neurodegenerative diseases. With age, the endothelium becomes less permeable for the transport controlled by membrane receptors or specific transporters. Therefore, essential proteins are less able to pass from the blood into the brain. In reverse, proteins that normally do not enter the brain in youth (e.g. albumin, fibrinogen or some auto-antibodies) are increased in brain tissue of old age and may trigger inflammatory processes. Moreover, the capillaries are less permeable to oxygen and sugar and the astrocyte processes around them change and become more plump. This reduces the transport of water by redistribution of specific water channels (known as aquaporins in the astrocyte membrane). The accompanying inflammatory component by activation of microglia cells additionally complicates the removal of no longer needed proteins.

In people with memory disorders in old age, the vascular changes are usually detectable long before the onset of symptoms. Interestingly, the vascular pathology in the brain can also be present if other organs are not affected by the narrowing and hardening of blood vessels (arteriosclerosis). Conversely, even if there is a coronary heart disease, the blood supply to the brain may be relatively normal. In summary, circulation disorders may occur organ-specifically and are to be seen as an important cause of aging in general and neurodegeneration in particular.

## In a Nutshell

- The more than 80 billion neurons of the brain are supplied by a network of blood vessels more than 600 km long.
- The perivascular Virchow-Robin space enlarges during sleep and thus leads to a faster removal of proteins no longer needed. In this way, the brain is "cleaned" at night.
- The blood-brain barrier prevents the uncontrolled passage of cells and larger molecules. It is primarily formed by the innermost layer in vessels, the endothelium.
- Via special transporters and membrane receptors in endothelial cells, sugar and other water-soluble molecules enter the brain. This process slows down with age.
- However, the level of certain proteins that normally do not enter the brain (e.g. auto-antibodies) is elevated in the aging brain and can trigger inflammatory processes.

## 2.2    Parkinson's Disease

Unlike during general aging, in neurodegenerative diseases distinct areas of our brain are affected in the early stages. These can be core regions such as neuronal nuclei, i.e. localized collections of nerve cells below the cerebral cortex or cerebral cortex areas, especially in the frontal and temporal lobes of our brain. The affected nuclei are typically characterized by the messenger substances they produce. In the second most common neurodegenerative disorder, Parkinson's disease, nerve cells that produce biogenic amines degenerate, in particular, dopamine (Fig. 2.5). Dopamine, like norepinephrine and adrenaline, belongs to the catecholamines, a subgroup of biogenic amines. Potentially toxic oxidants are formed from amines via monoamine oxidase, an enzyme that possibly plays an important role in the degeneration of nerve cells in aging and disease.

The syndrome first described in 1817 by James Parkinson is found in 2% of people over 55 years of age. In many countries (but not everywhere), it occurs more frequently in men than in women. In 2030, there will be about 9 million people affected by the disease worldwide. In 2020, about 400,000 people in Germany were suffering from Parkinson's disease. Each year, 12,500 patients are newly diagnosed. The disease is characterized by a typical symptomatology characterized by stiff and slow movements. A lack of movements (akinesia) is combined with muscle stiffness (rigor). Later on an expressionless, mask-like face becomes apparent.

Tying a loop or closing a button can become a serious problem. Often patients also have difficulty starting a particular movement, for example they cannot simply start walking. This is called *freezing*, that is a sudden and complete immobility. In addition, rhythmic muscle twitching such as the "pill-twisting tremor" of the hands (tremor), which is relatively slowly with a frequency of 4–6 Hz (Fig. 2.6) and may also affect the feet, can be found in four out of five patients. The combination of limited movement and tremor has led to the former name *Paralysis agitans*, although the disease is actually not a true paralysis. The innervation of the muscles, starting from the motor cortex, is intact. It is initially the neuronal nuclei located below the cortex that are affected. The survival rate of patients is 10–20 years after diagnosis.

### 2.2.1    General Pathomechanisms

The decisive problem of Parkinson's patients is that the amount of the messenger substance dopamine that is released in the subcortical motor areas, in the basal ganglia, is too low. The basal ganglia are indispensable for intended

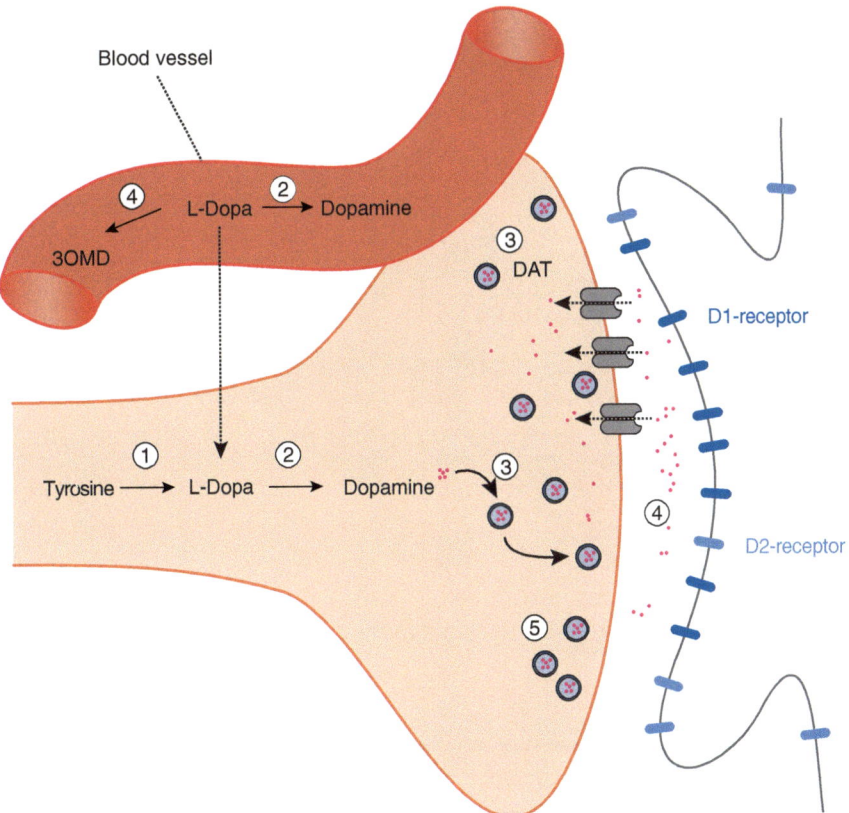

**Fig. 2.5**  The precursor of dopamine, L-Dopa, is formed by the rate-limiting enzyme of catecholamine biosynthesis, tyrosine hydroxylase, from the amino acid tyrosine (1). The Dopa-decarboxylase (2) then converts L-Dopa to dopamine, which is transported into small synaptic vesicles via specific dopamine transporters (DAT, 3). Such transporters are also found on the presynaptic membrane, from which dopamine (red dots) is reabsorbed after activation. In the area of the synaptic cleft, but also in the blood, L-Dopa is degraded by the catechol-O-methyltransferase (COMT, 4) to 3-O-methyl-dopa (3OMD). The monoamine oxidase of type B (MAOB, 5) is located on the outer mitochondrial membrane and inactivates dopamine to dopac (dioxyphenylacetic acid)

and involuntary movements. The striatum, the main nucleus of the basal ganglia, forms important nodes in modulatory neuronal networks that regulate and control the start and course of a movement. The dendritic processes of the striatal neurons retract in the absence of dopamine and the number of dopamine receptors decreases. Together, these changes result in the symptom triad akinesia, rigor and tremor as described above.

**Fig. 2.6** A Parkinson's patient leaning forward with a mask-like face and an implied tremor (iStock.com/LCOSMO)

The simultaneous deficiency of other amino acids, e.g. of serotonin or noradrenaline, leads to additional symptoms such as anxiety and sleep disorders. Patients may also complain of digestive disorders (mostly constipation), fatigue, muscle tension, a depressive mood or loss of smell. These problems often appear years before the motor symptoms of the disease. In the later stages, Parkinson's dementia develops in 20–40% of patients.

At the beginning of the disease, a general brain atrophy is not yet apparent, since initially circumscribed areas in the brainstem only are degenerating. The death of neurons located in the black-pigmented substance of the midbrain, known as Substantia nigra, is typical for Parkinson's disease (Fig. 2.7). Cells located in the lateral part of the pars compacta mainly supply the dorsal striatum with dopamine (the adjacent iron-containing pars reticulata of the substantia nigra is not affected).

Dopamine is released briefly and in high concentration at the synapses in the striatum in order to activate the D1 receptors present on the dendrites of the local neurons (Fig. 2.5). The D2 receptors are also found outside synapses. They are activated by the continuous release of low dopamine concentrations. D1 and D2 receptors are thus distributed differently on the neurons of the striatum (there are a total of five different dopamine receptors).

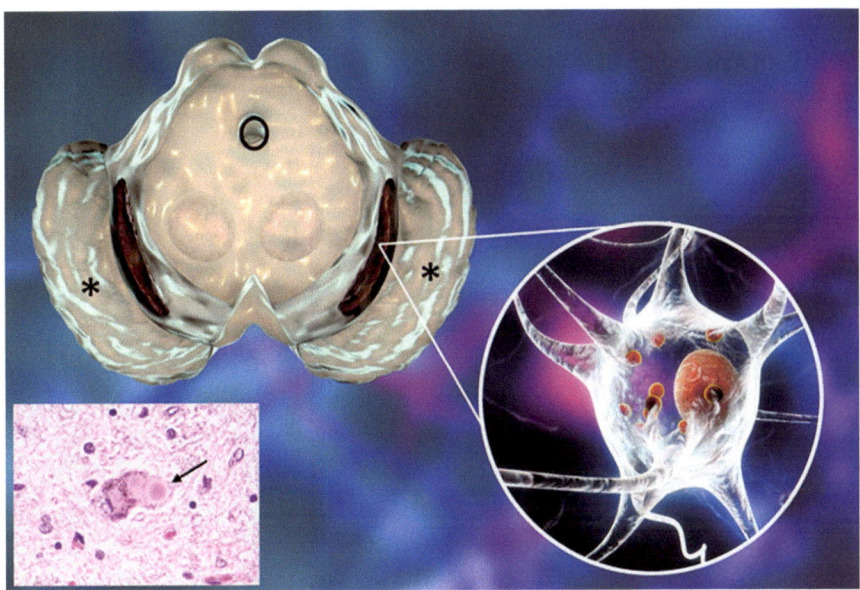

**Fig. 2.7** Cross section through the midbrain at the level of the substantia nigra. A dopaminergic neuron from the pars compacta is magnified. The cell body is shown with the nucleus and some endosomes (red). The "black substance" (substantia nigra) is located directly behind the cerebral peduncles (marked with a star), through which numerous connections run between the neocortex of the forebrain and the motor neurons in the brainstem and in the spinal cord. In the dorsal part of the midbrain a fine canal is observed, the aqueduct, through which the cerebrospinal fluid flows. In Parkinson's disease, the typical Lewy bodies (inset, arrow), first described by Friedrich Lewy in 1912, are detected within the substantia nigra neurons. Histologically, these represent cytoplasmic inclusion bodies with an eosinophilic (violet) nucleus (iStock. com/Dr_Microbe and Paulus et al., Neuropathology, Springer, Fig. 8.1a)

In addition, dopamine receptors are coupled to different signal transduction mechanisms and thereby influence the neuronal activation in opposite ways: D1 receptors bind an activating G protein, which stimulates adenylyl cyclase, while D2 receptors interact with an inhibitory G protein, which inhibits adenylyl cyclase. The effects of a neurotransmitter depend on the receptor for this particular transmitter. Nature achieves a large number of different effects by using a large variety of transmitter receptors, although it only uses a few transmitters. This explains why drugs that influence neurotransmission need to interact with specific receptors for a given messenger substance in order to achieve the desired effects.

The opposing symptoms of Parkinson's disease (on the one hand too little motor activity as in akinesia, on the other hand too much as in rigor and tremor) can thus be explained by dopamine affecting not only different neuronal networks but also acting on different dopamine receptors. In the upper striatum, the medium spiny neurons predominate, which express mainly D1 receptors. They are involved in the "direct" pathway and also project back to the substantia nigra. The nerve cells primarily carrying D2 receptors, on the other hand, are involved in the "indirect" motor pathways as shown in Fig. 2.8.

## The Problem with Parkinson's Begins in the Lower Brainstem

As discussed above, in Parkinson's disease a prominent cell loss is found in the nigro-striatal system with a significant reduction in dopaminergic signal transduction in the dorsal striatum (Fig. 2.9). In addition, however, a larger number of neurons is lost in other nuclei of the brainstem as well. These are involved in somatomotor systems but also in the activity of internal organs, i.e. in the visceromotor system. Interestingly, some of these brainstem areas are not only affected in Parkinson's disease, but also early on in Alzheimer's disease (see below). They include the locus coeruleus (noradrenaline) and raphe nuclei (serotonine) in the middle of the brainstem as well as the pedunculopontine nucleus with its cholinergic nerve cells and the dorsal vagus nucleus in the medulla oblongata. Disease-associated nuclei further include the histamine producing nucleus tuberomamillaris in the diencephalon and the cholinergic nucleus basalis in the forebrain.

The above mentioned nuclei are located in phylogenetically very old regions. Some of them are of great importance because they form especially long and highly branched axons in primates in order to supply the basal ganglia and the cortex, which are rapidly increasing in size during evolution. Their morphology and their intense cellular metabolism make these cells primary candidates for degeneration.

The enormous increase in the number of neurons in our cortex and in the striatum result in significantly more connections between the brainstem and the various target areas in the forebrain. Unfortunately, the brainstem nuclei cannot react with a comparable increase in the number of nerve cells due to the spatial restrictions in the posterior cranial fossa (*evolutionary constraints*). These neurons, therefore, have the only option to massively expand their axonal tree in order to meet the demands for synaptic transmission in the higher areas. They are metabolically more challenged than other nerve cells with fewer axon collaterals and succumb earlier to cellular stress.

Interestingly, the pathological changes in the cortex are first observed in those areas that have undergone significant structural changes over the course of development due to increasing functional requirements (form

**a**

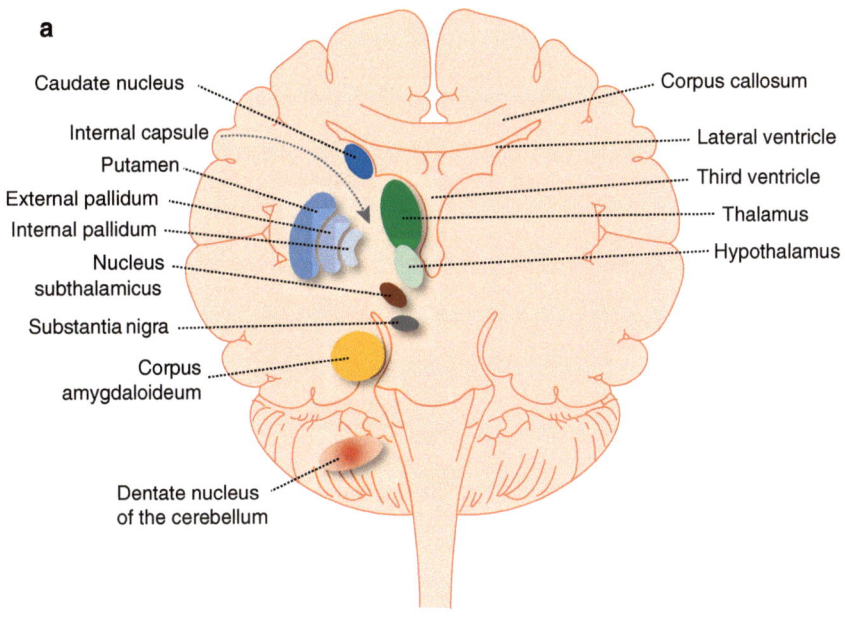

Caudate nucleus

Internal capsule

Putamen

External pallidum

Internal pallidum

Nucleus
subthalamicus

Substantia nigra

Corpus
amygdaloideum

Dentate nucleus
of the cerebellum

Corpus callosum

Lateral ventricle

Third ventricle

Thalamus

Hypothalamus

**b**

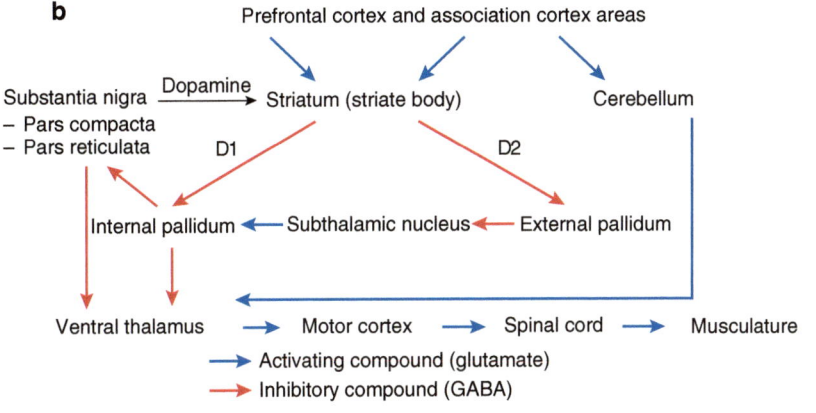

Prefrontal cortex and association cortex areas

Substantia nigra
– Pars compacta
– Pars reticulata

Dopamine

Striatum (striate body)

Cerebellum

D1

D2

Internal pallidum ← Subthalamic nucleus ← External pallidum

Ventral thalamus → Motor cortex → Spinal cord → Musculature

→ Activating compound (glutamate)

→ Inhibitory compound (GABA)

▶ **Fig. 2.8  a** View of a frontal section through the brain at the level of the basal ganglia. The lenticular nucleus (putamen and pallidum), the caudate nucleus and the amygdala are located in the forebrain. Putamen and caudate nucleus together form the striatum. The corpus callosum connects the two hemispheres of the brain. The inner part of the pallidum and the nucleus subthalamicus originate from the diencephalon, whose essential core nuclei are the thalamus and the hypothalamus. The substantia nigra is located in the midbrain and still part of the brainstem. It projects mainly to the dorsal striatum. The cerebellum with its largest nucleus, the nucleus dentatus, lies at the back of the brainstem. The nucleus dentatus sends the signals to the motor nuclei of the thalamus (in particular, to the nucleus ventrolateralis). From the ventral thalamus the signals reach the motor cortex areas and from there they project via the internal capsule (arrow) into the brainstem and the spinal cord. **b** Control of sensomotor functions by the central nervous system (CNS): The cortex stimulates, via activating transmitters (glutamate), neurons in the striatum that express the dopamine receptors D1 and D2. Via D1 receptors, dopamine transported from the substantia nigra into the striatum directly stimulates the motor programs (by double use of GABA, gamma-aminobutyric acid, as a transmitter, so ultimately activation results from disinhibition). After activation of D2 receptors, motor activities are regulated indirectly. The pallidum and striatum also project back to the substantia nigra (via inhibitory transmitters). The pars compacta is located dorsally to the pars reticulata of the substantia nigra (the latter sends inhibitory connections directly to the thalamus). The cerebellum acts in parallel to the basal ganglia circuits. However, the final part of the motor program from the thalamic nuclei to the motor cortex and further on to motor neurons in the brainstem and spinal cord is the same for both, the basal ganglia loop and the cerebellar loop

follows function). The entorhinal cortex located in the anterior temporal lobe would be such a region that lies directly adjacent to the hippocampus and plays an important role in memory formation as it is considered to generate the major input for the hippocampus. I will discuss this aspect in more detail in the chapter about Alzheimer's disease.

---

**In a Nutshell**

- Parkinson's disease is characterized by stiff and slow movements (rigor and akinesia). In addition, trembling in the hands or feet (tremor) is observed.
- These symptoms are caused by a lack of dopamine in the dorsal striatum, the main nucleus of the basal ganglia located below the cortex cerebri. Dopamine ist necessary for the control of motor programs stored here.
- Dopaminergic neurons are predominantly located in the substantia nigra of the midbrain.
- Dopamine is released from axonal endings and binds to dendrites of striatal neurons that carry different types of dopamine receptors (mainly D1 or D2).
- The substantia nigra and other brainstem nuclei affected in patients are phylogenetically very old and innervate disproportionately large target areas in the forebrain. These neurons therefore have particularly long and highly branched axons, the metabolic supply of which is demanding.

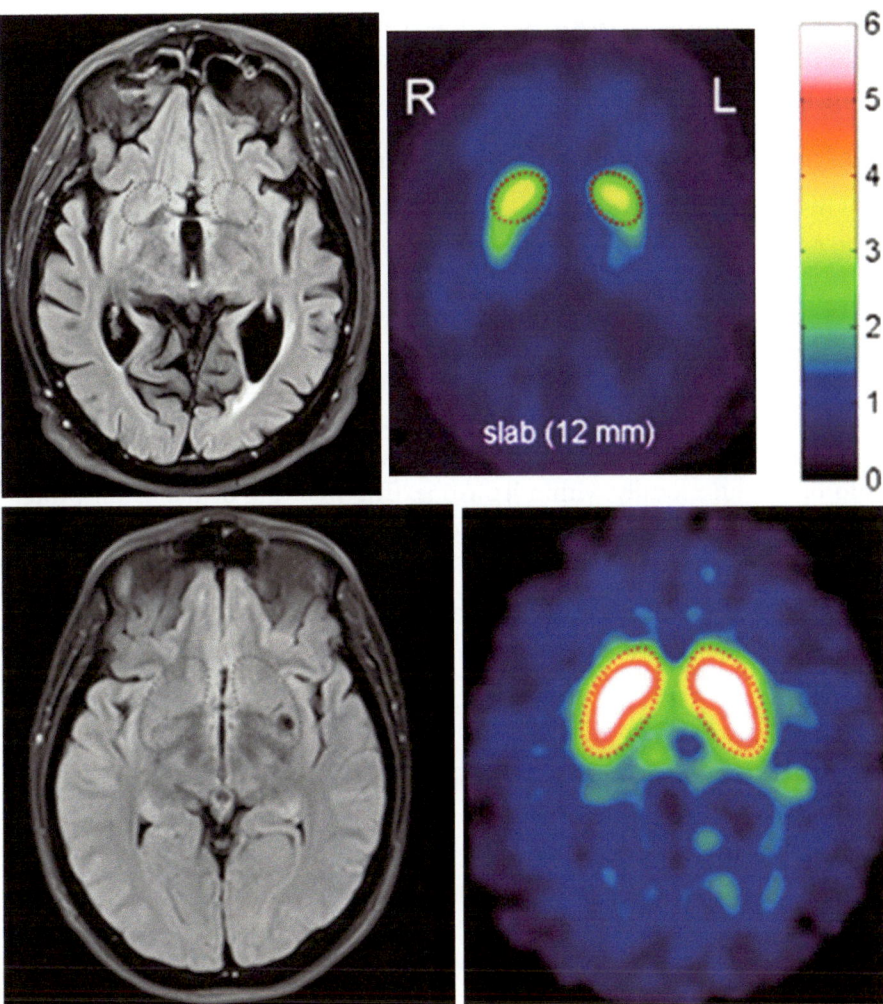

**Fig. 2.9** A horizontal (transverse) FLAIR (dark-fluid T2) MRI with depiction of dopamine receptors (right, in different colors depending on the concentration of receptors) using DAT-PET-CT (thanks to J. Fiehler, University Hospital Hamburg-Eppendorf, Germany). The upper part shows the brain of a 67-year-old Parkinson's patient with dementia. There are only a few dopamine receptors left in the basal ganglia. The lower image part shows a section through the brain of a 72-year-old without Parkinson's syndrome. The color intensity is shifted to red/white, which indicates a significantly stronger binding of the radiolabeled marker of dopamine receptors

## 2.2.2  Special Morphology of Affected Neurons

Early on in Parkinson's disease, nerve cells start to degenerate that innervate a large territorium in the forebrain, and therefore require complex axonal trees. These are composed of thin, poorly insulated (myelinated) axonal extensions that project from different brainstem nuclei into the basal ganglia and also into the cerebral cortex. Each individual axonal tree can reach a total length of over four meters due to its extensive branching. Moreover, up to two million synaptic contacts are assumed for each dopaminergic neuron innervating striatal neurons

These synapses contain between 300,000 and one million proteins (with about 10,000 different proteins per neuron). This results in 1–2 trillion proteins in the maximum configuration that a neuron would have to transport into or produce locally within its axons. The resulting demands on protein production and transport are enormous and require a lot of energy. In contrast, the average number of contacts on projection neurons in the cortex is comparably low (approx. 10,000). In fact, cortical pyramidal neurons are—like the smaller interneurons with short axons—generally not affected in patients with Parkinson's disease.

Neurons in the brainstem that project into the forebrain must therefore be equipped with a large number of mitochondria, endosomal vesicles, and cytoskeletal and membrane proteins due to the very high protein and energy demands. They are particularly sensitive to disruptions in the cellular metabolism and work constantly at their limit. We will discuss in more detail below why defects in the bioenergetic processes may cause Parkinson's or other neurodegenerative diseases.

Taken together, the scientific findings of the recent years have led to the hypothesis that various neurodegenerative disorders are ultimately based on disruptions in the metabolic balance of nerve cells that have been at their limits for many years due to their morphological complexity. If, in addition, defective proteins are produced by these cells, as in the monogenic disease forms, the built-in safety mechanisms are no longer sufficient to protect these neurons from early cell death. The function of defective proteins often cannot be compensated by intact proteins so the modulatory networks including these neurons will fail. At that time point, clinical symptoms will become visible.

In an early stage of Parkinson's disease, only the axons of affected neurons are damaged. The corresponding cell bodies in the substantia nigra and in other aminergic nuclei shrink, but they are still alive. This is evidenced by the fact that degenerative changes in the striatum can already be detected in Parkinson's patients years before the death of the perikarya, which corresponds to the loss of their distal axons. Axonal degeneration occurs through the proteases already mentioned, the caspases, but also through calpain, a calcium-dependent protease.

Cellular damage due to very high bioenergetic demands is now generally accepted as major cause of neurodegeneration, which is further accelerated by exogenous influences, e.g. toxins. This hypothesis is also relevant for Alzheimer's disease (see Sect. 2.3) and amyotrophic lateral sclerosis (ALS). The latter is a severe muscle paralysis caused by loss of motor neurons in the cortex, brainstem and spinal cord. Here, the topic just discussed repeats itself: a single motor neuron in the anterior horn of the spinal cord has a long and highly branched axon that supplies up to 1000 muscle fibers. Therefore, some neurons have to maintain axonal trees of several meters in length here, analogous to the dopaminergic neurons of the substantia nigra with their extensive axonal ramifications.

## Special Requirements for Highly Branched Neurons

To cover the enormous biological demands to maintain meter long axons, the necessary proteins are produced not only in the cell body, but also at free ribosomes in the axon and in the synapses themselves. Special transport systems have to be provided by the nerve cell, which deliver protein-coding matrices, the respective mRNAs, into the neuronal processes. It is therefore not surprising that in patients suffering from neurodegenerative diseases mutations in genes coding for mRNA-binding proteins have been found repeatedly in genetic studies. These proteins are essential to transport mRNAs from their place of origin in the neuronal nucleus to the ribosomes in dendrites and axons.

If mRNA transport in nerve cells is defective, the local production of proteins required for the maintenance of axons and their endings is disturbed. This prevents the synapse from fulfilling its physiological function, i.e. releasing defined amounts of neurotransmitters. In addition, a large number of well-functioning mitochondria is required for this task, as the energy requirement is naturally high in long axons. The cellular protein and lipid machinery runs at full capacity and long transport distances have to be covered (one molecule of ATP is required for 8 nm of transport distance in the axon).

A number of safety mechanisms have therefore been developed due to the high metabolic demands particularly in neurons. These arise intracellularly, but also systemically with regard to the formation of neurons in excess during brain development. Because of this neuronal back-up the loss of a few hundred nerve cells per day does not stand out. In neuromodulatory networks, which form the basal ganglia dependent motor activity circuits, even thousands of neurons can fail before any clinical symptoms occur. We suspect that, in Parkinson's disease, around half of the 500,000 pigmented neurons in the pars compacta of the substantia nigra have already died by the time the symptoms set in. The disease is therefore nearly always diagnosed at an advanced stage (from the perspective of a neuropathologist).

Another reason for the relatively late onset of clinical symptoms in neurodegenerative diseases is found in the compensatory mechanisms that reinforce parallel-connected neural networks, so that our brain can still perform its multitude of tasks for a long period of time. This ability is of particular importance when considering the fact that we practically do not form any new neurons and no long-distance axonal connections between the different brain regions after completing development at the age of around 20.

## Many Amine-releasing Neurons are Constantly Active and Therefore More Easily Stressed

In addition to the long, branching axonal projections, dopaminergic and noradrenergic brainstem neurons show a pacemaker function, that is, an intrinsic activity of the cells without exogenous stimulation. This ensures that a certain amount of messenger substance is permanently available in the target area. The dopaminergic nerve cells of the substantia nigra therefore emit electrical signals (they spike) independent of their synaptic input at a frequency of 2–10 Hz. The resulting calcium fluctuations, mediated by voltage-dependent $Cav1\text{-}Ca^{2+}$ channels in the plasma membrane, represent another risk factor for neuronal degeneration, especially when the calcium buffer capacity is limited.

The entry of calcium into mitochondria is a requirement for ATP synthesis and thus to ensure the energy supply of neuromodulatory networks. This is an evolutionary very old mechanism that can also be found in muscle cells. So, if there is a temporary ATP deficiency, the activity of the cell does not come to a standstill. That would be a real danger, because otherwise all targeted movements would lose their drive in the event of an acute power failure, e.g. at times of glucose deficiency. Such permanent activity can also be observed in other essential neuronal circuits, such as in those responsible for the formation of a circadian day/night rhythm, but also for the fear-or-flight reaction following activation of the sympathetic nervous system. All of these situations require the provision of sufficient energy carriers. It is therefore not surprising that the energy consumption of the central nervous system is very high, especially within synaptic contacts.

But everything has its price and so the mitochondria in neurons suffer over time. Constantly increased calcium concentrations results in increased permeability and hyperpolarization of the mitochondrial membranes, which in turn leads to increased release of oxygen radicals (ROS) and reactive nitrogen species (RNS). In addition, calcium stimulates the protease calpain mentioned above and thus promotes the aggregation and toxicity of α-synuclein, a key protein in Parkinson's disease. Hence, clinical studies with calcium channel blockers (e.g. dihydropyridines) have their justification not only in cardiology, but also in Parkinson's patients.

Taken together, nerve cells in the substantia nigra and in other brainstem nuclei with highly branched axonal processes must be considered particularly vulnerable due to their morphological and electrophysiological peculiarities in old age (Fig. 2.10). It is not surprising that oxidative damage occurs in these cells from the age of 60 onwards. Moreover, mitochondrial and endosomal transport disorders preferentially lead to neuronal degeneration in evolutionarily old brainstem areas.

**In a Nutshell**

- Long and highly branched axonal processes pose high demands on protein and mRNA transport as well as on the energy supply with ATP.
- Genetic changes in mitochondrial or mRNA-transporting proteins therefore predispose to Parkinson's disease.
- The pacemaker function of the nigro-striatal neurons affected in Parkinson's disease ensures that a certain amount of dopamine is always available in the target area.
- The continuous activity leads to calcium fluctuations, which in turn cause increased permeability of mitochondrial membranes and intracellular release of harmful oxygen radicals (ROS) and reactive nitrogen species (RNS).
- In Parkinson's disease about half of the 500,000 dopaminergic neurons of the pars compacta of the substantia nigra have to die before clinical motor symptoms occur.

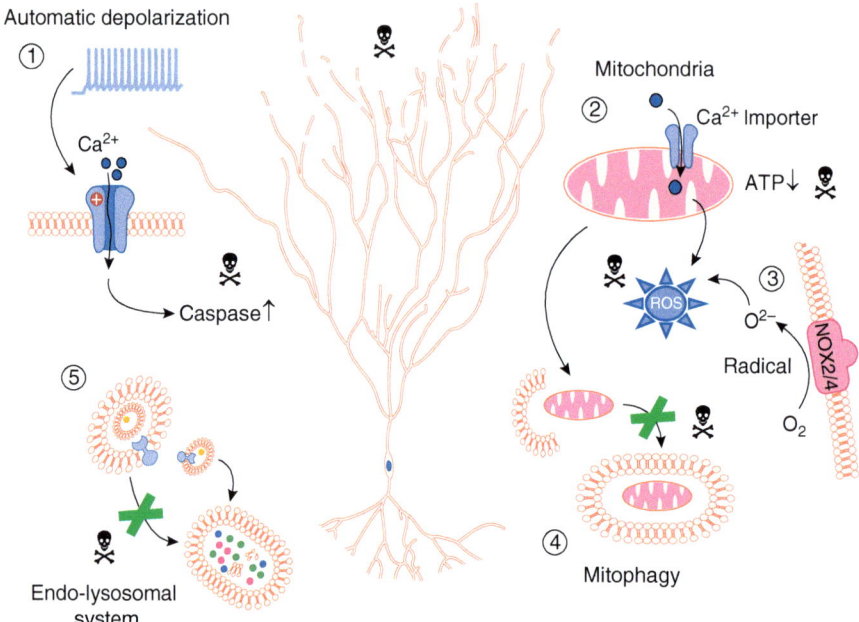

**Fig. 2.10** Summary of the dangers to which dopaminergic neurons in the substantia nigra and their highly branched axons in the striatum are exposed. The automatic depolarization (1) leads to increased calcium influx into the cell, which can activate caspases and damage mitochondria (2). Free radicals, which are formed in mitochondria and by membrane-bound enzymes (3), lead to oxidative stress and are therefore potentially toxic. Genetically determined defects in mitophagy and autophagy (4) as well as in the endo-lysosomal system (5) disturb protein homeostasis and result in axonal degeneration and finally in premature neuronal cell death

## 2.2.3  Specific Causes of Parkinson's Disease

A secondary Parkinson's disease can be caused, for example, by drugs that have an anti-dopaminergic effect through blockade of dopamine receptors. Haloperidol would be such a substance that is used in delusions and other psychoses. In addition, circulatory problems can also cause a Parkinson's syndrome. In contrast, the sporadic (non-familial) form of Morbus Parkinson is a form of primary Parkinson's disease. Previously, environmental toxins were blamed for causing the symptoms (pesticides, herbicides or metals), and traumatic brain injuries were said to increase the likelihood of developing the disease. In fact, Parkinson's disease is sometimes diagnosed in professional boxers at middle age, including in former heavyweight boxing champion Muhammad Ali.

Although it is generally accepted that exogenous influences lead to an increased incidence, we assume that primarily intrinsic (endogenous) risk factors ultimately cause the disease. However, external influences may anticipate the onset of symptoms. On the other hand, there are also indications that environmental factors delay the onset of the symptoms. For example, a statistically significant reduction in the risk of Parkinson's has been shown to be associated with regular consumption of coffee. This is attributed to polymorphisms (changes in the DNA base sequence) in a gene (CYP1A2) that is important for the metabolism of caffeine. Moreover, individual DNA changes have been described, e.g. the Q65P mutation in a lysosomal ion channel (LysoK$_{GF}$), that improve the function of this particular channel and thus reduce the risk of developing symptoms. However, the majority of the genetic abnormalities found are associated with an increased risk of Parkinson's disease.

As discussed above, it is nowadays generally accepted that a strong metabolic stress on nerve cells in the brainstem and unfavorable genetic predispositions must come together to trigger neuronal degeneration. The brainstem nuclei mentioned above are restricted in various functions many years before the disease begins. They reveal changes in protein homeostasis, for example. In particular, defects in protein degradation can be detected, which after a longer time interval lead to an accumulation and aggregation of selected proteins in the cell. Disturbances in the mitochondria, which result in reduced energy production, can also be observed.

The fact that many patients have at least one first-degree relative who also suffers from the disease speaks for a significant role of genetic factors in the pathogenesis of neurodegenerative disorders. In 5–10% of all Parkinson's patients, an inherited cause of the disease can be expected. Some of these genes have been identified in recent years by special DNA analysis of the entire genome including all 20,000 protein-coding genes.

These studies are called **G**enom **W**ide **A**ssociations **S**tudies (GWAS), which are carried out using modern next-generation sequencing methods in several centers worldwide in order to include as many patients as possible. Interestingly, in such genetic analyses, individual base changes (single nucleotide polymorphisms, SNPs) are occasionally found, which affect some of the already well known familial Parkinson's disease associated genes. Sometimes these are located in DNA sequences that do not directly code for the relevant proteins, but are associated with them, for example, by influencing the transcription of their mRNAs.

It was previously assumed that only a few younger patients show a monogenic form, i.e. a predisposition to Parkinson's disease caused by a single gene change. However, it is possible nowadays to actually detect such a gene defect in about half of patients under 50 years of age. Importantly, inherited gene mutations that lead to the production of defective proteins are present in all cells (and not only in certain neuronal nuclei). This means that it is unlikely that they are solely responsible for neuronal degeneration. In contrast, an additional pathomechanism must be present in affected nerve cells, for example, the metabolic requirements underlying highly branched axons as described above.

Today we know that more than 20 different genes can cause Parkinson's disease in their mutated form. One of these is the gene coding for the ubiquitin ligase parkin, which is necessary for the mitophagy i.e. the degradation of old or damaged mitochondria by autophagy (Fig. 2.3). An inactivating mutation disrupts this function of parkin and, consequently, mitochondrial defects are to be expected. Similarly, PINK1 (PTEN-induced putative kinase 1), an enzyme required for mitophagy, is associated with a hereditary form of Parkinson's disease. The normal turnover of mitochondria is apparently of crucial importance for neurons that are highly metabolically and bioenergetically demanding. As mentioned above, disruptions in energy supply lead to premature aging (senescence) or cell death. It seems that a reduced activity of mitochondrial complex I, the NADH dehydrogenase, is primarily responsible for this process to occur.

**In a Nutshell**

- There is no known, specific cause for primary (idiopathic) Parkinson's syndrome.
- Circulatory disturbances or medications that block dopamine receptors may result in secondary Parkinson's disease.
- The disease is not caused by environmental toxins or traumatic brain injury, but can occur earlier as a result of these.
- In 5–10% of all Parkinson's patients (in 50% of those under 50 years of age), an inherited cause of the disease is detected.
- More than 20 genes have been discovered so far that, in their mutated form, lead to Parkinson's disease. Some of them affect the assembly and disassembly of mitochondria.

## 2.2.4  Alpha-synuclein: A Key Protein in Parkinson's Disease

The best-known Parkinson-associated gene is SNCA and was first discovered in its mutated form in a large Italian family in 1997. The mutation is inherited in an autosomal-dominant manner and leads to a structural change in the protein encoded by SNCA, α-synuclein (the homologous β-synuclein is encoded by the SNCB gene). In affected patients molecularly fine, fibrous structures, the α-synuclein fibrils, with an diameter of about 10 nm are found.

Moreover, α-synuclein binds to various other proteins. One of them is ubiquitin, which acts as a molecular marker for proteins to be degraded in all cells as described above. Intracellular aggregates containing synuclein, ubiquitin and other proteins are referred to as Lewy-bodies after the neurologist Friedrich Lewy (1885–1950). They are complex mixtures of proteins, lipids, cellular membranes and deformed organelles, i.e. mitochondria or lysosomes can be found in these aggregates, too (Fig. 2.7). The importance of α-synuclein in the context of Parkinson's disease is shown by the fact that, in addition to a degeneration of the substantia nigra, the detection of an accumulation of this protein is sufficient for a definitive diagnosis of idiopathic Parkinson's disease.

Normal α-synuclein is a soluble, 140 amino acid long and 14 kilodalton (kDa) heavy protein. With around 1% of the total cytosolic protein amount, it is one of the most common proteins in neurons. It binds to phospholipids in plasma membranes by interactions of its N-terminal domain located at the beginning of the amino acid chain. This amphipathic (simultaneously hydrophilic and lipophilic) section of α-synuclein consists of eleven amino acids, which are repeated four times. The resulting spatial secondary structure corresponds to an α-helix that can bind to other α-helices and membranes. Hence, α-synuclein plays an important role in the binding of vesicles thereby regulating the exocytosis of dopamine and other transmitters. In fact, synuclein interacts with the SNARE protein complex, which is responsible for the fusion of vesicle membranes with the outer plasma membrane.

In addition, two α-synuclein molecules can bind to each other via their central NAC domain (non-amyloid component) and form a dimer. High concentrations of α-synuclein or a single amino acid exchange within the molecule, for example in an SNCA mutation, leads to the formation of oligomers comprising 2 to 100 individual monomers. However, Lewy aggregates can also be formed by post-translational modifications of α-synuclein.

These include phosphorylation, ubiquitination, nitration, but also trunca-tion of the protein, which promote oligomerization and fibril formation.

The production of these molecular fibers takes place through pathologi-cal phase transitions that are determined by changes in the physico-chemical properties of the oligomers and other molecules involved. If the number of molecular interactions exceeds a critical threshold, there is a sudden change in the specific characteristics of these biomolecular condensates (oligomers) (Fig. 2.11).

Physical characteristics include the viscosity, elasticity or surface tension of the condensates floating in the cytoplasm. Genetic variation, post-transla-tional modification, pH changes in the cell or oxidative processes, e.g. by the HNE mentioned above binding to α-synuclein, promote their formation. In addition to proteins, they contain various RNA molecules and are finally converted into an insoluble aggregate, the above mentioned Lewy body, in a liquid-to-solid phase transition.

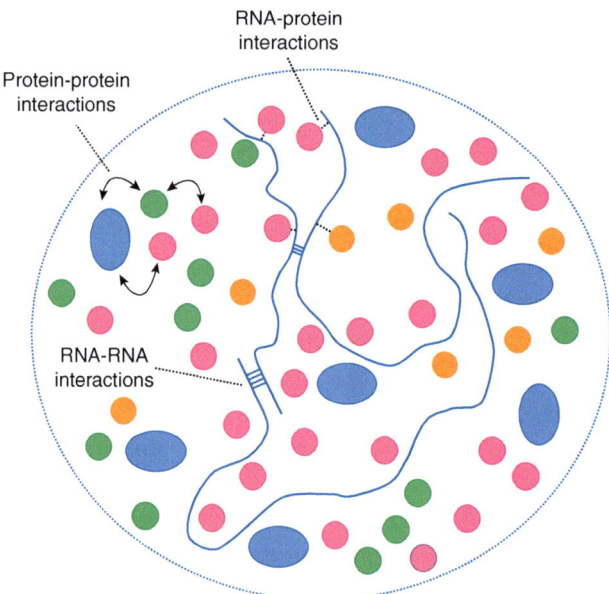

**Fig. 2.11** After a phase transition, biomolecular condensates form in the neuronal or glial cytoplasm. They represent molecular islands of special physico-chemical proper-ties in which various interactions between individual proteins, protein oligomers and RNA molecules occur. The exact pathomechanistic significance of these condensates is still unclear

## Fibrils Are More Dangerous Than Aggregates

According to this theory, it is not the insoluble aggregates, but the α-synuclein fibrils that are the actually toxic intermediate stages on the way to condensates, since they block the release of transmitters by binding to synaptic vesicles and attach to membranes of various organelles (mitochondria, lysosomes, ER, Golgi apparatus) and to the proteasome. In this way, the fibrils significantly impair the functions of key organelles. The fibrils may even penetrate and destroy the plasma membrane directly. Moreover, after being secreted from the cell, they activate a Toll-like receptor (TLR2 and TLR4) on microglia. This triggers an inflammatory reaction in the nervous tissue that can accelerate the neurodegenerative processes. In addition, α-synuclein binds to a microglial Fc receptor (FcγRIIB) and can thus influence the phagocytic activity of the cells which also affects degenerating neurons, i.e. microglia can completely remove neurons that are not yet dead.

However, the α-synuclein released by nerve cells not only binds to microglia, but is also taken up by astrocytes and can form aggregates in these cells—just as in neurons. The astroglia responds with an induction of pro-inflammatory cytokines (IL-1, IL-6 and TNFα) and chemokines (CXCL1). Again, a Toll-like receptor (TLR4) is required for this. Finally, the uptake of glutamate by astrocytes and their involvement in the blood-brain barrier is disturbed. Interestingly, an increased expression of α-synuclein mutants (A53T) in astrocytes also leads to the degeneration of nerve cells with corresponding motor deficits in animal models corroborating the importance of glial cells in the pathogenesis of Parkinson's disease.

It is thus of particular importance for aging nerve and glial cells to be able to control the concentration of α-synuclein intracellularly. For this purpose, the lysosomal autophagy system (especially the chaperone-mediated autophagy mentioned above), but also the proteasome, are required. However, if autophagy is defective or only partially active in older neurons, α-synuclein and other proteins will further accumulate. In such a vicious circle, α-synuclein oligomers thus lead to an additional inhibition of intracellular protein degradation. Not surprisingly, it was shown that reducing α-synuclein levels in the brain can significantly improve motor deficits in animal models of the disease.

Interestingly, however, α-synuclein is not only a substrate for autophagy, but can also impair the formation of autophagosomes. Attempts to stimulate autophagy pharmacologically, e.g. with the drug rapamycin, have so far failed due to the immunological side effects (rapamycin suppresses the immune system). However, it has been possible to stimulate neuronal autophagy using genetically modified astrocytes which reduced the neurotoxicity of α-synuclein by taking up neuronal organelles and protein aggregates that have been secreted by nerve cells.

## Aggregates Are Not Only Found in Nerve Cells

Since an increase in α-synuclein is also observed in other neurological diseases, Parkinson's disease is referred to as a synucleinopathy. These disorders include dementia with Lewy bodies or multiple system atrophy (MSA). The former is characterized by α-synuclein aggregates in the frontal cortex. In MSA, the aggregates are mainly found in oligodendrocytes, that is, in the glia forming the myelin.

It must be repeated that in neurodegenerative diseases, the described genetic variations are found in all body cells, and the consequences of a mutation for cellular metabolism are in principle relevant for glial cells as well. However, there are different amounts of the relevant gene products in different cells, so their expression levels are usually different. For example, α-synuclein is formed to a much lesser extent in astrocytes than in neurons.

An increase in the normal, non-mutated α-synuclein can be caused by duplications or triplications of the SNCA gene or by single base exchanges in the non-coding sections of the DNA, so-called enhancer regions. A mutation in the SNCA promoter, for example, could lead to increased production of the protein through increased transcription of the mRNA. The degree of overexpression of α-synuclein mRNA correlates with the severity of the disease in terms of a more rapid and combined disease course with additional symptoms, such as dementia.

### In a Nutshell

- The best-known mutation leading to Parkinson's disease involves α-synuclein, a protein that is highly expressed in neurons and tends to oligomerize, form fibrils and finally aggregates (known as Lewy bodies).
- Alpha-synuclein oligomers and fibrils block neurotransmitter release and damage plasma membranes which result in neuronal degeneration.
- The release of α-synuclein leads to the activation of Toll-like receptors on microglia and to its uptake by astrocytes. Both of these can trigger an inflammatory reaction in the brain.
- The more α-synuclein is produced, the more severe is the disease.

## 2.2.5 The Prion Theory of Parkinson's Disease

Pathological α-synuclein can also be taken up from outside into neurons and passed on to adjacent neurons via synaptic contacts. This results in spreading of the Parkinson pathology in the brain. Exogenously applied antibodies against α-synuclein inhibit the transport of α-synuclein from nerve cell to nerve cell. Interestingly, however, the transfer of the protein does not take place on all synaptic contacts, because some neurons seem to be more resistant to such a "synuclein infection" than others.

Proteins can be discharged directly out of the cell via the outer plasma membrane or they are excreted from vesicles having different names: microvesicles, exosomes or apoptotic bodies. They are generally classified according to their size, their secretory pathway and specific marker proteins. Exosomes arise—in contrast to microvesicles—from larger vesicles, the multi-vesicular bodies (MVBs, see Fig. 2.3).

MVBs fuse with the plasma membrane and release their contents, including exosomes, into the extracellular space. Exosomes are therefore small vesicles (50–100 nm in diameter) and made visible by electron microscopy or high-resolution light microscopy. If the lysosomal system is no longer fully functional, α-synuclein oligomers are transferred in larger quantities to adjacent neurons via exosomes and spread throughout the brain.

The transfer of α-synuclein oligomers from pathologically altered cells to adjacent healthy cells, which then acquire the same pathology, follows the prion theory first described by Stanley Prusiner. It states that externally introduced, improperly folded proteins may cause a neurodegenerative disease by practically forcing the same, correctly folded proteins in nerve or glial cells to adopt the misfolded, pathological structure.

This hypothesis was first described in humans in the context of kuru disease in Papua New Guinea and further developed in the 1980s in the context of scrapie disease in sheep and of bovine encephalitis (BSE). Thus, there is a transfer of a quasi "infectious" agent without viruses, bacteria or fungi being involved. Incorrectly folded prion proteins have been detected in the form of fibrils and amyloid deposits in the brains of patients suffering from Creutzfeldt-Jakob disease, a rapidly progressive form of dementia.

**Cranial Nerves Transport Pathological Proteins into the Brain**

Transmissible, prion-like properties are discussed for α-synuclein, but also for the Aβ peptides and for the Tau protein which are increased in

Alzheimer's disease and described in the next section. Interestingly, the typical Alzheimer's deposits and Lewy bodies are often found together in the brain. In particular, both occur early on in the olfactory bulb, which could lead to smell disorders described in both diseases.

The olfactory bulb has numerous axonal connections to the olfactory epithelium in the nasal cavity, from which foreign substances and toxins, but also inflammatory mediators or viruses can easily enter the brain. It is striking that protein deposits are already detectable early on not only in the olfactory system, but also in the brainstem with the nucleus dorsalis nervi vagi particularly affected.

The dorsal vagal nucleus forms the origin of the efferent, intestine-regulating visceromotor part of the tenth cranial nerve, the Nervus vagus. It connects the brainstem directly with the organs in the chest and abdominal cavity. The tenth cranial nerve is thus in principle able to take up pathogenic substances such as viruses, toxins or prions in the mucous membranes of thoracic and abdominal organs and transport them retrogradely to the brainstem. Key aspects of this research work and the classification of Parkinson's disease used today in neuropathology go back to the well-known German anatomist Heiko Braak (Fig. 2.12).

Various animal experiments seem to confirm Braak's hypothesis. For example, pathological α-synuclein is not found in the vagal nucleus when the vagus nerve is cut before injection of α-synuclein into the intestine. Intriguingly, Lewy pathology has been found in intestinal biopsies of Parkinson's patients. The digestive disorders described before the onset of typical Parkinson's symptoms may be explained by this observation. An example of Lewy body-like aggregates in the nerve plexuses of the intestine is shown in Fig. 2.13.

From the intestine, pathological proteins could thus reach the medulla oblongata and then continue on to those brain areas that are synaptically connected to the visceromotor neurons of the dorsal nucleus of the vagus nerve ("gut-brain axis"). Although definitive confirmation of this theory is still pending, it provides a possible explanation for the development of neurodegenerative diseases by pathological changes outside the brain. In addition, pathological α-synuclein has also been found in the skin of Parkinson's patients, so that a simple skin biopsy could support diagnostics in the future.

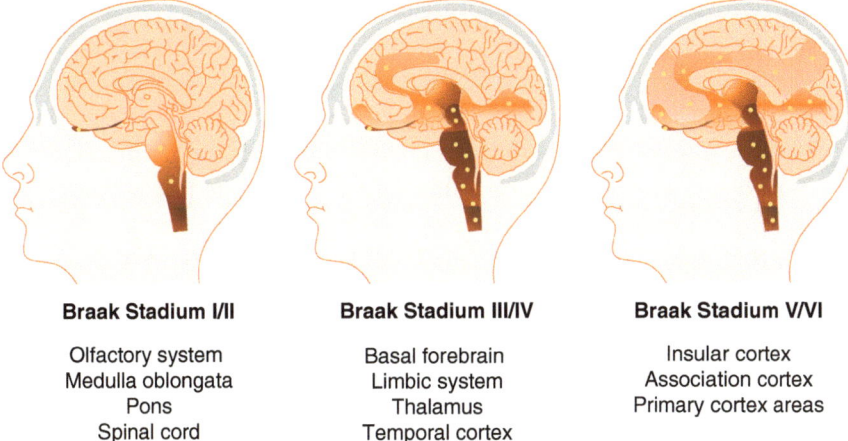

**Braak Stadium I/II**

Olfactory system
Medulla oblongata
Pons
Spinal cord

**Braak Stadium III/IV**

Basal forebrain
Limbic system
Thalamus
Temporal cortex

**Braak Stadium V/VI**

Insular cortex
Association cortex
Primary cortex areas

**Fig. 2.12** In early stages of Parkinson's disease (Braak stages I/II), intracellular α-synuclein aggregates (yellow dots) are found preferentially in aminergic and cholinergic nuclei of the brainstem (dark brown). In the following stages (III and IV), neurons in the midbrain and the basal forebrain are infiltrated, possibly via a prion-like mechanism. In the advanced stages (V/VI), neuronal degeneration and Lewy body pathology are also found in the substantia nigra and the neocortex. In this scheme, the superior sagittal sinus is grey-marked. It is a large blood vessel formed by the dura mater through which venous blood from the neocortex and a part of the cerebrospinal fluid with degradation products of Lewy aggregates is drained.

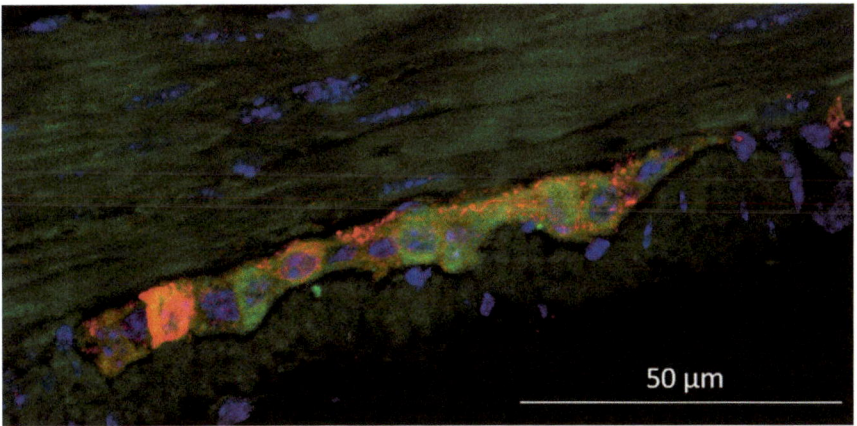

**Fig. 2.13** An enteric ganglion located between the two muscle layers of the human large intestine with red-marked nerve cells and light green fluorescent synuclein aggregates (K.H. Schäfer, Molecular Neurodegeneration, DOI 10.21203/rs.3.rs-86154/v1)

## The Difficulties of a Clear Pathogenesis of Parkinson's Disease

It has been suggested that enteric macrophages are involved in the formation of pathological aggregates in the intestine. It is therefore assumed that the immune system plays an important role in the intestine-brain axis. In this context, the importance of the intestinal content, the microbiome, must be discussed in relation to our brain. We know that a variety of bacteria, but also viruses and fungi, in the small and large intestine can influence our nervous and immune system throughout the body. This effect likely occurs via fatty acids and lipopolysaccharides that are released by the intestinal microorganisms and act on the bloodstream or via nerve connections on the brain. Based on observations of mice that are kept bacteria-free in special cages, it must be assumed nowadays that our microbiome deteriorates the pathology of neurodegenerative diseases. For example, it has been observed that a general reduction in the number of germs in the intestine, e.g. by administration of antibiotics, can positively influence the clinical symptoms in Parkinson's and in Alzheimer's disease.

It is therefore possible that the neurodegenerative pathology starts in the intestine (or elsewhere in the peripheral nervous system) and is transfered from cell to cell via prion-like mechanisms. However, the sole cause of Parkinson's disease cannot come through the vagus nerve into the brainstem, because the probability of the disease is not increased in people whose tenth cranial nerve has been surgically divided after gastric or intestinal ulcers. Interestingly, an appendectomy, that is, the surgical removal of the vermiform appendix, is associated with a reduced risk of Parkinson's disease.

We have to conclude that the exact Lewy pathology remains unclear as yet. The trigger is still missing, if there is **one** trigger only at all. Furthermore, it is not clear why Lewy bodies occur frequently in brain regions where nerve cells rarely degenerate. They are not only formed in the dopaminergic substantia nigra, but also in various other nuclei of the brainstem, especially in those that produce biogenic amines (adrenaline, noradrenaline and serotonin). Finally, these aggregates are not specific to Parkinson's disease, as they also occur in other neurological diseases. Surprisingly, Lewy bodies are only occasionally found in areas connected by axons to the substantia nigra or to the locus coeruleus, even though they should be passed on to these nuclei by synaptic contacts according to the prion hypothesis. It can therefore be assumed that the formation of the aggregates requires special intrinsic properties of the affected neurons that we do not yet all know.

Finally, the most common (autosomal-dominant) hereditary forms of Parkinson's disease, namely mutations in the LRRK2 gene (Leucine-rich repeat kinase 2), do not always lead to the formation of Lewy bodies, even though autophagy and the degradation of α-synuclein are also disturbed. In patients with Parkin mutation, Lewy bodies also occur less frequently than in sporadic forms. Taken together, not all patients reveal the typical Parkinson pathology in the brain, and vice versa, the pathology also occurs in people without the disease. It is therefore assumed that Lewy bodies are neither necessary nor sufficient for neuronal cell death in the brains of Parkinson's patients and do not correlate with the severity of the clinical symptoms. Hence, we are in a difficult situation scientifically. Probably a variety of pathogenic mechanisms lead to the clinical picture of Parkinson's disease, just as different causes lead to neuronal degeneration in the different forms of dementia, which will be discussed in the next section.

---

**In a Nutshell**

- α-Synuclein and other proteins can be released from the cell via exocytosis. They are packed into small vesicles (exosomes) and secreted.
- The pathological forms of α-synuclein thereby spread throughout the brain and damage other cells (according to the prion theory).
- In addition to the olfactory nerve, the vagus nerve may be involved in the transmission of pathogenic substances (viruses, toxins, prions) from the periphery to the CNS.
- However, Lewy bodies are also detected in brain regions where nerve cells rarely degenerate. Since they are found in several neurological disorders, they are not specific for Parkinson's disease.

---

## 2.3    Dementia and Alzheimer's Disease

In Germany, more than 1.5 million people are considered demented, i.e. 9% of the population over 65 is affected. In addition to the individual suffering, dementia is also of social importance because the global costs exceed 500 billion euros per year. The risk of dementia doubles approximately every 5 years from the age of 60. Age is thus the most important risk factor for the development of dementia. More than 50 different forms can be found in medical literature, including the disease first described by Alois Alzheimer in 1906, which accounts for 60–70% of cases. Almost half of those over 85 years of age have dementia, two thirds of them are women. This is probably due to the decline in the female sex hormone estrogen after menopause, as sex hormones also have neurotrophic effects in the brain, especially in the hippocampus, by stimulating metabolism and securing network stability. Their absence can therefore accelerate the death of neurons. On the other hand, recent evidence suggests that the rise in plasma level of the follicle-stimulating hormone (FSH) around menopause may play a key role in the development of dementia in women.

About 10% of Alzheimer's patients experience an early form of the disease between the ages of 40 and 50. Most patients live an average of 7–8 years after diagnosis, but in some cases it can be 20 years. The disease is characterized by a slow course and a progressive decline in the number of nerve cells in many regions of the brain. This leads to a reduction in the functional reserve over several years. An exception to this rule is the rare Creutzfeldt-Jakob disease, which progresses rapidly and usually leads to death within a year.

The vascular dementia caused by insufficient blood supply is the second most common form and often begins quickly, but then remains stable for years before the symptoms worsen in episodes. The problems depend on which brain region is most affected by the circulatory problems. Patients often have a history of heart attacks, strokes or diabetes mellitus. Due to the vascular pathology present in most people over the age of 65, the vascular component is suspected to play an important role in Alzheimer's disease, too. Therefore, we assume a mixed form of both types of dementia in at least 50% of the patients.

Finally, in a small number of those affected, the disease apparently also results from repeated head injuries and the resulting encephalopathy. Although prospective studies on this important question still need to be conducted, the available data suggest that, for example, professional footballers due to numerous headers and collisions are three times more likely to develop dementia in old age than other athletes who do not suffer such head injuries.

## 2.3.1  How Does Alzheimer's Disease Manifest Itself?

The first episodes of memory problems often become apparent through spatial disorientation, which points to degenerative changes in the caudal (ventral) parts of the temporal lobe. Patients also forget names for ordinary objects, such as household appliances. In the clock test, it is difficult for them to correctly enter the numbers 1 to 12 in a given circle. Later, all three explicit memory functions, i.e. the acquisition, long-term storage and retrieval of new information, are significantly impaired. Names, telephone numbers and conversation contents are easily forgotten. In addition, word finding difficulties and extensive paraphrasing are observed.

Unfortunately, the loss of long-term memory also results in the loss of reference to one's own life story. Impressions from childhood remain the longest accessible. In a late stage of the disease, patients are no longer able to handle financial or administrative matters. Their ability to criticize and to judge is significantly restricted. In addition, delusions can occur. In contrast to amnesia, the pure memory loss, and aphasia, the pure language disorder, in Alzheimer's dementia, several cognitive functions are affected at the same time. Neuropathologically, a reduced (atrophic) brain with enlarged lateral ventricles and cortical sulci as well as significantly narrowed gyri are visible (Fig. 2.14).

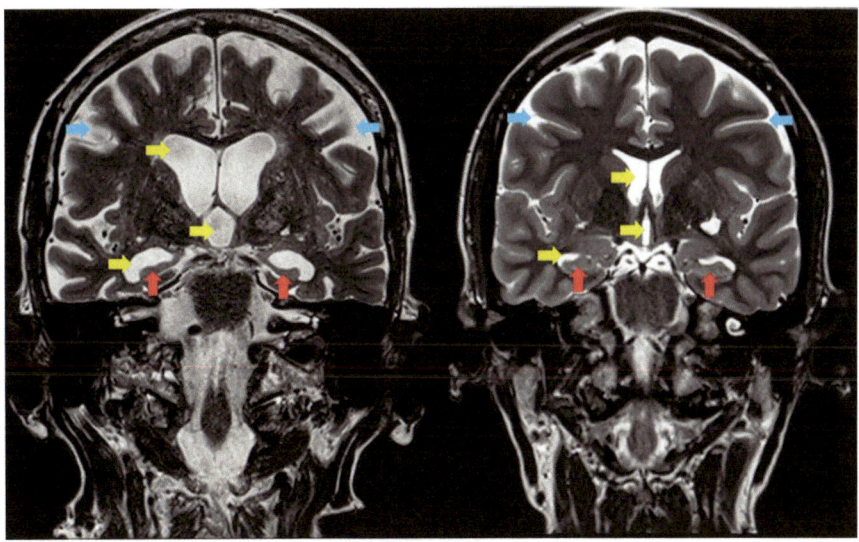

**Fig. 2.14** In this frontal section through the brain of a 75-year-old patient with high-grade Alzheimer's dementia (left), significantly enlarged ventricles (yellow arrows) and outer CSF spaces (blue arrows) can be seen compared to a 72-year-old patient without cognitive impairment (right). Note the atrophied hippocampus (red arrows) in the temporal lobe of the Alzheimer patient. The cerebrospinal fluid (CSF) is white in this (T2-weighted) MRI, the cortex is light gray and the white matter is dark gray (J. Fiehler, University Hospital Hamburg-Eppendorf, Germany)

**In a Nutshell**
- The risk of dementia doubles every 5 years after the age of 60.
- More than half of all dementia patients suffer from Alzheimer's disease. Women are more likely to be affected than men.
- A lack of blood flow to the brain may result in vascular dementia. A combination of both forms, Alzheimer's and vascular, dementia is common.
- Spatial disorientation indicates a problem in the temporal lobe.
- In the further course of the disease, all three explicit memory functions, that is, the acquisition, storage and retrieval of new information, are impaired.

## 2.3.2  General Pathomechanisms

Neuropathologically, early deficits in glucose metabolism and changes in neurotransmitter production, predominantly in the cholinergic system, can be detected. The most relevant nucleus for synthesis of the neurotransmitter acetylcholine in the forebrain, the nucleus basalis, atrophies. Cholinergic neurons in the septum are also affected, which is of particular importance for deficient hippocampal functions. The cholinergic supply of the frontal cortex, the temporal lobe and the gyrus cinguli above the corpus callosum is particularly impaired. Since these areas play an important role in our explicit memory their cholinergic supply is of great importance for the understanding of Alzheimer's disease (Fig. 2.15).

As mentioned above, most Alzheimer's patients are over 65 years old and have no family history of dementia, i.e. their relatives are not affected by Alzheimer's dementia. However, after carrying out large-scale genetic analyses, special features are detected also in some of the sporadic cases, such as an increased occurrence of the lipoprotein ApoE4 (encoded on chromosome 19). In addition, environmental factors are assumed to act as disease triggers in 35% of Alzheimer's patients. These can be very different and range from a shortened school education to circulatory disorders and social deprivation (loneliness) or hearing loss. However, in 60–70% of cases, a genetic variation is assumed as underlying pathomechanism.

The classic Alzheimer's genes include the amyloid precursor protein APP located on chromosome 21, as well as presenilin-1 on chromosome 14 and presenilin-2 on chromosome 1. Mutations of these genes can cause an early onset dementia and are usually inherited in an autosomal-dominant fashion, i.e. half of the children of an affected person pass the mutation on to the next generation. In addition, gene duplications, single base exchanges and

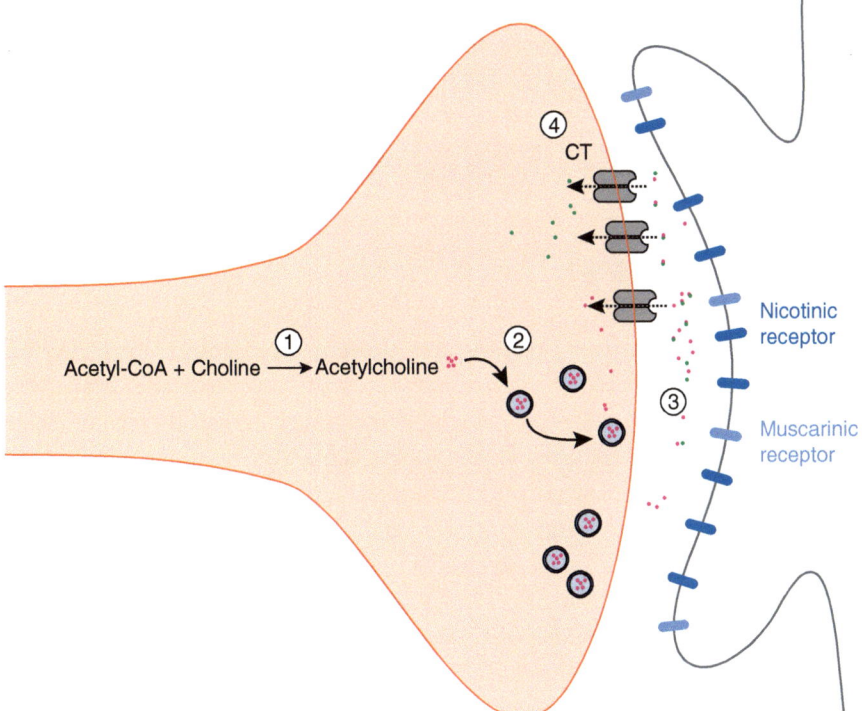

**Fig. 2.15** Acetylcholine is formed by the enzyme choline acetyltransferase (1) from acetyl-CoA (generated in the mitochondria) and choline. Acetylcholine is then taken up from the cytosol into neurosecretory vesicles via a vesicular acetylcholine transporter (VAchT) (2). Up to 10,000 acetylcholine molecules are contained in each vesicle. After exocytosis and diffusion across the synaptic cleft, acetylcholine activates ion channels (nicotinic receptors) or G-protein-coupled (muscarinic) receptors, which are named after their respective agonists (nicotine or muscarine). The enzyme acetylcholinesterase (3) cleaves the transmitter (red dots) back into choline and acetic acid (acetate), thus terminating the effect of acetylcholine at the synapse. A choline transporter (CT, 4) in the pre-synaptic membrane leads to the re-uptake of choline (green dots) into the axonal terminal, as nerve cells cannot synthesize choline themselves

mutations are involved. Genes that affect the innate immune system or lipid metabolism can also be affected.

Essential for Alzheimer's disease—as in Parkinson's disease—is a neuronal degeneration that is associated with the formation of typical protein deposits. Two forms of aggregates are distinguished: the senile amyloid plaques and intracellular fibrils. The latter are different from the α-synuclein fibrils discussed above, occur in the later course of the disease only and, in contrast to the plaques, correlate well with existing memory disorders.

Amyloid plaques can be detected with Amyloid Positron Emission Tomography (PET) and consist ultrastructurally, i.e. at high magnification in the electron microscope, of fibril bundles 5–10 nm in diameter. Since they occur as part of normal aging processes in around half of all older people, they are not specific to the disease and do not in themselves cause memory deficits.

## The Amyloid Pathology

In senile plaques, various proteins and protein fragments (called amyloid) are deposited (Fig. 2.16). Degenerated neuronal processes and signs of an inflammatory process characterized by numerous microglial cells are found in the vicinity of the plaques. It has long been assumed that they represent the actual cause of neuronal degeneration in Alzheimer's disease. This idea became known as "amyloid hypothesis" or "β-amyloid cascade" in the early 1990s through the work of Hardy and Higgins.

Today it is assumed that the plaques arise compensatory as a consequence of pathologically high concentrations of soluble protein fragments that are rendered ineffective and thus harmless by aggregation (similar to the α-synuclein and Lewy body pathology discussed above). In any case, the amyloid plaques are not specific for Alzheimer's disease and can also occur in large numbers in cognitively intact, 100-year-old people.

The senile plaques are mainly formed by Aβ (beta-amyloid). These are 39–43 amino acids long peptides that arise from a precursor protein, APP (amyloid precursor protein). Increased concentrations of Aβ peptides can also be found in various other neurological diseases, e.g. in traumatic brain injury, ischemia, multiple sclerosis and even after prolonged anesthesia or as a consequence of sleep deprivation. They are preferably detected intracellularly in the endo- and lysosomal compartment, may be released into the extracellular space and removed via the perivascular spaces or degraded enzymatically by neprilysin or insulin-degrading enzyme (IDE). Importantly, Aβ peptides can cross the blood-brain barrier.

APP is a type 1 membrane-associated glycoprotein that is localized in particular to vesicles, but also found in blood platelets (thrombocytes). Since the APP gene is located on chromosome 21, the corresponding protein product is increased in humans with Down syndrome (trisomy 21, i.e. threefold occurrence of chromosome 21). Interestingly, Alzheimer's dementia occurs in Down patients at an average of 52 years, much earlier than

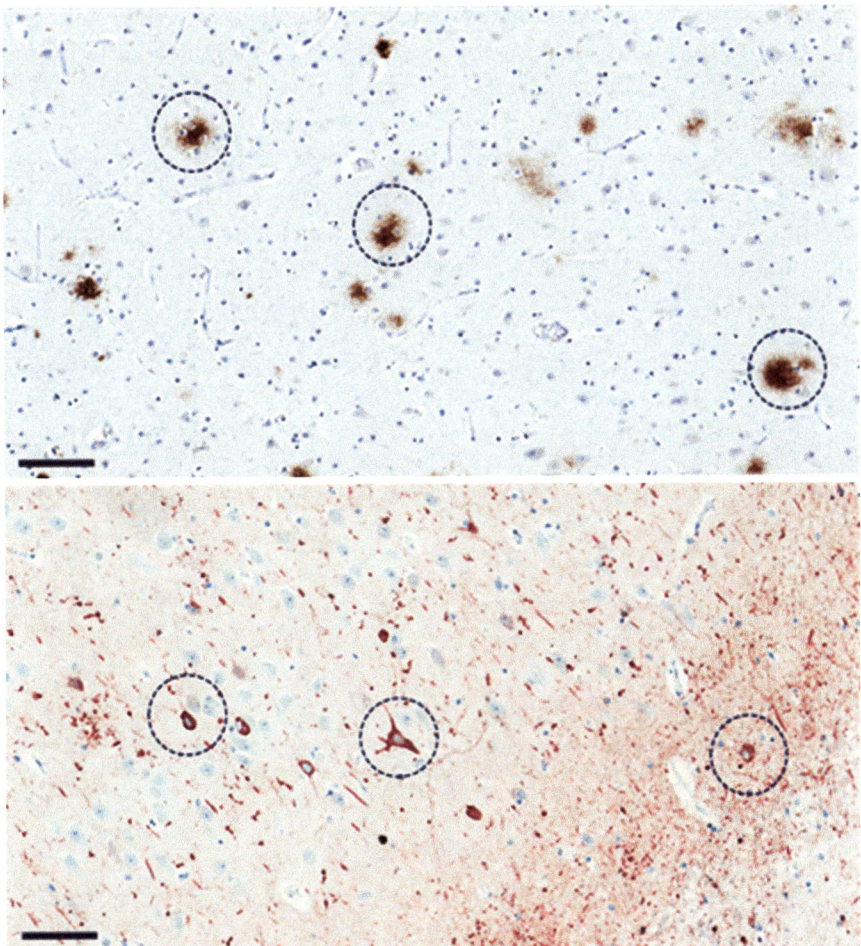

**Fig. 2.16** Immunohistochemical staining of histological sections from the brain of Alzheimer's patients. The brown amyloid plaques located extracellularly are marked in the upper image, and the red Tau fibrils located intracellularly are encircled in the lower image. Tau fibrils are observed not only in neuronal cell bodies, but also in axonal projections. Neuronal and glial cell nuclei are stained blue. The bar in the images corresponds to 100 μm (A.M. Birkl-Töglhofer, Med. Uni. Graz)

in the general population. By the age of 65, almost 90% of all trisomy 21 patients are suffering from dementia. It is therefore likely that APP plays an important role in the pathogenesis of Alzheimer's disease, possibly by amplifying the above-mentioned ER stress.

A processing of APP requires the activity of membrane-bound proteases (secretases). The enzyme α-secretase cleaves the APP domain located within the membrane (a process called ectodomain-shedding) and thus prevents the formation of Aβ peptide (Fig. 2.17). A water-soluble product (secreted APP or sAPP) is formed, which can even have favorable, protective effects like a neurotrophic factor. The proteases responsible for the formation of Aβ-peptide are β-secretase (also called BACE1) and γ-secretase. The latter contains presenilin-1 (PS1) or presenilin-2 (PS2) as catalytic subunit and cleaves the Aβ-peptides from the transmembrane fragment of APP. Mutations in the presenilin genes therefore lead to a changed composition of the APP cleavage products. There is an accumulation of the longer, less soluble Aβ-peptides, which dimerize (forming pairs) or oligomerize. The latter

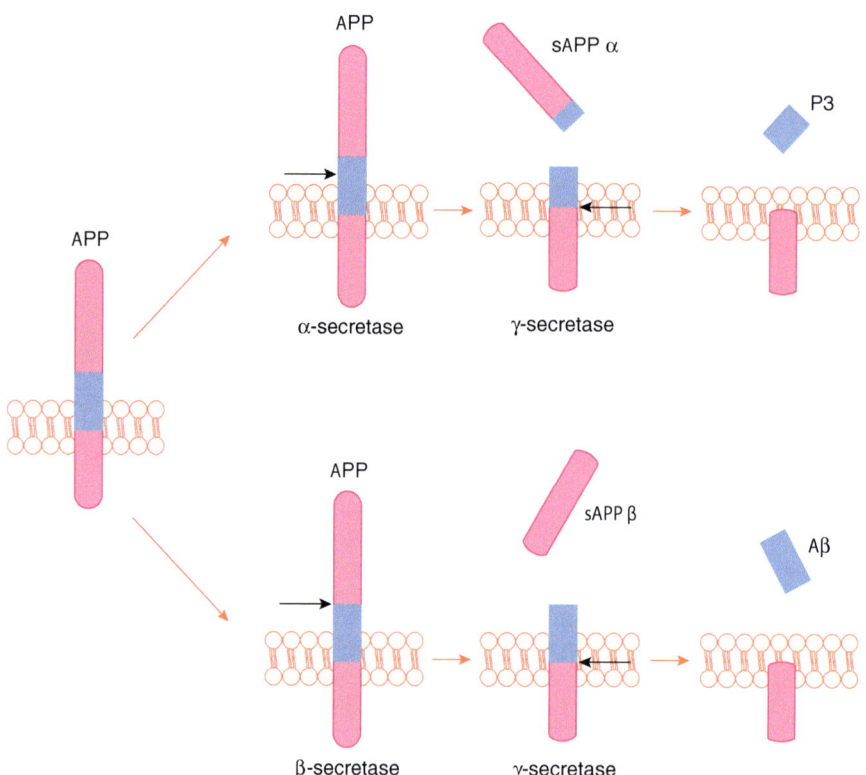

**Fig. 2.17** The membrane-bound APP is cleaved by three proteases (α-, β- and γ-secretase). Cleavage by the α-secretase releases the secreted APPα (sAPPα). The subsequent degradation of the membrane-bound peptide chain by the γ-secretase leads to the P3 peptide, which does not form stable oligomers that could be toxic. In Alzheimer's disease, there is an increased activation of the β-secretase, which cleaves the APP to form sAPPβ. The following activity of the γ-secretase causes the formation of Aβ (after T. Grune, Aging Processes and Neurodegeneration, Fig. 4.1, Springer)

are likely the primarily toxic forms for neurons (analogous to the α-synuclein pathology described above).

Mutations in the APP or in the two presenilin genes account for about 5% of Alzheimer's cases. But mutations that can prevent or at least delay the onset of Alzheimer's dementia can also be found. These include, for example, the exchange of alanine for threonine at position 637 of the APP gene (right next to the β-secretase interface). This change is also called Iceland mutation (Ala673Thr) because of its locally high occurrence in Iceland. It reduces the occurrence of amyloid plaques by 40–50%. Other mutations, e.g. at position 22 (Gln22Glu), are located in the Aβ region. These show a pronounced formation of amyloid plaques, but no fibril pathology with dementia. A slightly altered structure of the Aβ region can therefore trigger the disease, but on the other hand block processes that lead to neuronal damage. It is generally accepted nowadays that APP and its cleavage products occupy key positions in the neuronal metabolism, from which they may decide the fate of the nerve cell in old age.

## In a Nutshell

- In Alzheimer's disease, cholinergic neurons degenerate mainly in the basal frontal lobe of the brain. Acetylcholine plays an important role in learning and memory.
- At the genetic level, carrier of the ApoE4 gene have a twofold increased risk of developing Alzheimer's disease.
- Mutations primarily affect the amyloid precursor protein (APP) gene on chromosome 21 and the two presenilin genes.
- Therefore, increased amounts of APP are observed in patients with Down syndrome (trisomy 21), who usually develop Alzheimer's dementia by the age of 50.
- Presenilin-1 and -2 form the catalytic subunits of γ-secretase, which is necessary for the production of Aβ.
- Longer, poorly soluble Aβ peptides oligomerize and are toxic to nerve cells.
- Aβ peptides are sequestered by the formation of large aggregates (β-amyloid plaques) and made harmless.
- Amyloid plaques are not specific to Alzheimer's disease and also occur in cognitively intact older people.

## The Many Effects of Aβ Peptides

The 42 amino acid long Aβ42 peptide and the two amino acids shorter Aβ40 have key functions in Alzheimer's disease. The longer Aβ peptides tend to pair in the form of a dimer or oligomer. They are then still soluble, but tend to aggregate. In addition to Aβ, other peptides prone to interaction are also present in Alzheimer's disease, e.g. Aeta-amyloid (amyloid-η). However, Aβ42 is the most commonly detected peptide in amyloid plaques, whereas Aβ40 accounts for around 60% of soluble peptides in the cerebrospinal fluid (followed by Aβ38 and Aβ42).

In particular, the Aβ40 peptide is important for the normal metabolism, as it can protect the brain from bacterial infections (presumably as part of our innate immunity). It is also involved in the repair of nerve injuries or blood-brain barrier disorders. So Aβ40 usually has important functions that even require oligomerization in terms of the antibacterial effect, because otherwise Aβ40 cannot bind to bacterial walls. In this respect, Aβ peptides behave similarly to other antimicrobial peptides that act as pore-forming toxins, i.e. they create holes in the bacterial wall. The toxic effect of Aβ oligomers in neurons may similarly be explained by binding to the plasma membranes of nerve cells. It has been shown that aggregated Aβ42 peptides can also form pores that allow calcium, sodium or cesium ions to enter the cell and thus become neurotoxic.

We have known for some time that Aβ peptides can also cause overexcitability of nerve cells, especially in those that have intrinsic (endogenous) spontaneous activity (as described above in Parkinson's disease). In addition, they change signal transduction cascades and cause errors in synaptic transmission. This disturbs the long-term potentiation at synapses, which is absolutely necessary for memory formation. In addition, the release of the activating messenger glutamate is stimulated, which is toxic for adjacent neurons in too high concentrations. Apparently, the synaptic glutamate receptor (NMDA) is not responsible for this, but a cation channel associated with the NMDA receptor (TRPM4).

Finally, negative effects of Aβ amyloid on blood supply have been described, since vessels become constricted and cerebrospinal fluid drainage impaired by a reduction in glymphatic activity in the perivascular space. On the other hand, glial cells, i.e. astrocytes and microglia, can bind and take up excess Aβ oligomers, which delays the formation of extracellular plaques.

## 2.3.3 The Disturbed Protein Homeostasis in Alzheimer's Disease

In the brain areas particularly affected by Alzheimer's disease, the activity of the proteasome, the cellular recycle bin, mentioned above is significantly reduced. This means that fewer proteins can be degraded. If protein synthesis remains constant, the cell will be overloaded with protein. Some proteins bind to each other, forming aggregates, and thus interfere with the normal flow of organelles in the cell. A better understanding of the intracellular protein transport and degradation system will therefore be key to understanding Alzheimer's disease.

The membrane-enclosed vesicles involved in intracellular transport, the endosomes, are essential for cell survival. In particular, in nerve cells with their large number of long processes, endosomes are of paramount importance for the release of transmitters from vesicles and for the provision of transmitter receptors at synapses. In recent years, neuropathologists have therefore directed their research interest to the endosomal system.

The early endosomes (see Fig. 3.3), which separate from the surrounding outer plasma membrane and migrate into the cell interior, are significantly enlarged in the brains of Alzheimer's patients. It has long been suspected that changes in the intracellular transport precede the Aβ peptide accumulation. It was also assumed that other peptides and proteins accumulate in nerve cells and damage them. All three cell biological mechanisms (endocytosis, intracellular trafficking and exocytosis as described in Fig. 2.3) must therefore function properly in order to avoid congestion in cellular protein transport. Neurons can withstand such blockades for a certain period of time only.

It is therefore likely that Alzheimer's disease and probably also other neurodegenerative diseases are mainly due to disturbances in protein transport and protein homeostasis, respectively (Fig. 2.2). In Alzheimer's disease, βCTF, a fragment of APP formed by β-secretase, impairs endosomal transport and thus leads to the accumulation of phosphorylated tau proteins, which are responsible for the formation of the intracellular fibrils to be discussed below.

As mentioned above, initial extra- and intracellular protein deposits may have a favorable effect. They reduce the amount of freely diffusing, probably toxic proteins by aggregation. Danger arises, in particular, when Aβ peptides or soluble Tau forms interfere with endosomal transport and thus cause congestion, which in turn leads to an increase in the peptides. Such a vicious circle can develop quickly. Deposition sites for the peptides would eventually be exhausted and the neuron would perish.

Thus, the mutations of the presenilin genes PS1 and PS2 mentioned above not only lead to an increase in Aβ, but also to a disturbance of the endo-lysosomal system. For example, a PS1 mutation prevents the necessary acidification of lysosomes and thus impairs autophagy, the disposal of large protein complexes and deformed organelles. Interestingly, some of the gene defects found in the above-mentioned GWAS studies are precisely associated with proteins that are involved in the early phase of autophagosome formation and membrane transport between endosomes. These include, for example, the GTPase Rab11 that is associated with recycling endosomes (see Fig. 3.3).

The exact molecular defects in cellular metabolism that ultimately lead to the accumulation of damaging substances and to the loss of nerve cells may be different for each patient, so that, with neurodegenerative diseases (similar to many tumors), only a personalized treatment tailored to each individual patient would help to prevent the disease from breaking out in the first place. Such a therapy needs to be **causal**, i.e. it would stop the degenerative process that causes the disease. The alternative would be to start at the end point of the disease-causing mechanisms, i.e. at neuronal cell death. Some experimental therapies already exist that can at least delay neuronal cell death. Finally, there is also the option of replacing damaged parts of the brain by new nerve cells, i.e. by stimulating neurogenesis. These strategies are discussed in the third chapter.

### In a Nutshell

- The Aβ40 peptide can have positive effects: it is antibacterial and promotes repair processes in the brain.
- Large amounts of Aβ42 peptides aggregate and lead to overexcitability of neurons by pore formation in membranes and release of the activating messenger glutamate, which is toxic in high concentrations.
- As in Parkinson's disease, protein transport and degradation are impaired in Alzheimer's dementia. APP fragments inhibit endosomal transport and autophagy.
- The exact pathomechanisms that ultimately lead to premature neurodegeneration and dementia are likely to be different in most patients.

## 2.3.4  Tau Pathology

In contrast to the amyloid plaques, which can be numerous without causing a cognitive deficit, the amount of Alzheimer's typical intraneuronal fibrils correlates well with the occurrence of memory problems. These fibrils, also called tangles, consist of highly phosphorylated and aggregated tau. Tau is a primarily axonal protein that can be detected by tracers in imaging techniques, for example by positron emission tomography, but also by antibodies in the cerebrospinal fluid. Phosphorylated tau (p-tau) means that numerous phosphate groups were enzymatically coupled to the protein. These phosphorylations can be attributed to the activity of three enzymes: Glycogen synthase kinase (GSK3β), protein kinase C (PKC) and protein kinase-N1 (PKN1).

However, an increased phosphorylation of tau could also be explained by a reduced activity of protein phosphatase 2A (PP2A), since this enzyme is responsible for the removal of phosphate groups. A large amount of coupled phosphate groups (hyper-phosphorylation) prevents—just like mutations in the tau protein—the normal degradation of tau and disturbs its binding to microtubules (Fig. 2.18). Both of these effects impair neuronal metabolism,

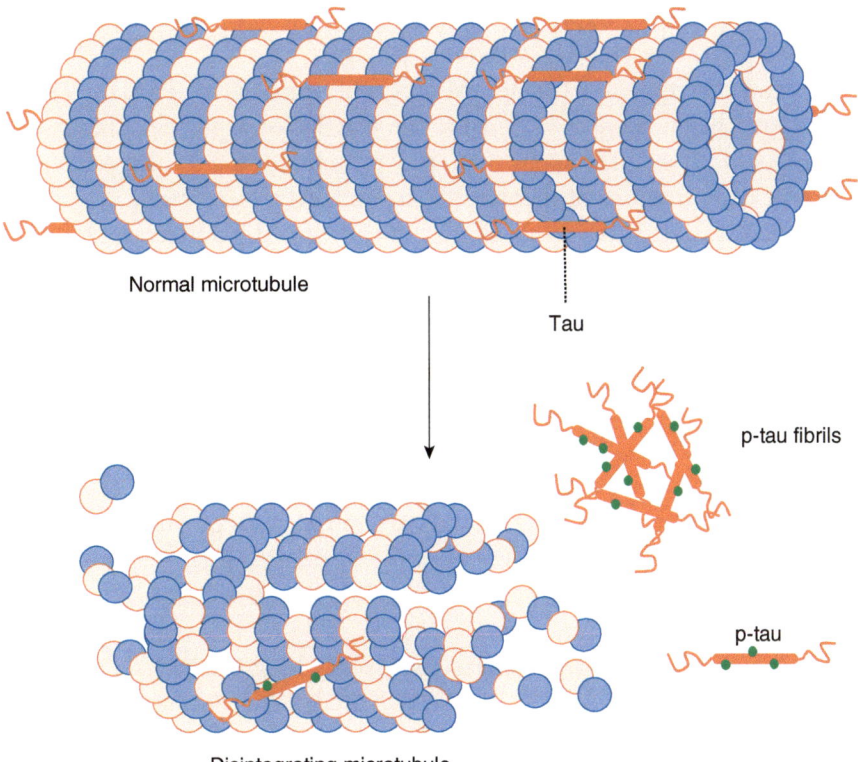

Normal microtubule

Tau

p-tau fibrils

p-tau

Disintegrating microtubule

**Fig. 2.18** Microtubules form long, intracellular tubes that are made up of individual α- und β- tubulin proteins. Tubules form the core of the cytoskeleton. They stabilize the neuronal cell body and especially its extensions, dendrites and the axon. Tau binds to microtubules and is essential for their assembly. Tau modifications, e.g. by coupling of phosphate groups (green dots), lead to the protein's detachment from the microtubules, which then depolymerize, i.e. the tubulin chains break down into their individual components (α- und β-monomers). Hyperphosphorylated tau (p-tau) tends to aggregate and forms the tau fibrils that occur more frequently in Alzheimer's disease

because tau is an important stabilizer of microtubules, which enable the directional transport of vesicles and form part of the neuronal cytoskeleton. They provide structure and shape to axons and dendrites. Newer findings suggest that tau can also bind to synaptic vesicles via a membrane protein, synaptogyrin-3, and disturb their function.

Just as the synucleinopathies arise from increased α-synuclein protein levels, tau deposits are characteristic of the tauopathies. These are disorders that are caused by a mutation of the tau gene. Examples would be frontotemporal dementia (FTLD), Pick's disease, corticobasal degeneration or progressive supranuclear palsy. In addition to Alzheimer's disease, other neurological diseases include chronic traumatic encephalopathy, which is associated with an intact tau gene, but a tendency to form tau aggregates.

The tau-coding MAPT-gene (**M**icrotubule-**A**ssociated **P**rotein **T**au) is located on chromosome 17. Usually, the matrix for a protein, the mRNA, is composed of several exons (i.e. the protein-coding DNA sequences). Different combinations of these exons in the MAPT-gene result in a total of six different versions (isoforms) of the tau protein. The different mRNAs are created by alternative splicing of the long pre-mRNA transcripts produced by RNA polymerase. The process refers to the selective cutting out of some exons or of non-coding DNA sequences, the introns.

The final tau mRNAs contain partial exon repetitions, e.g. from exon10 (so that the 3R form of tau contains three exon repetitions). Tau proteins are then linked to each other via such *tandem repeats* and thus aggregate into fibrils. This explains why some of the so-called splicing factors, i.e. proteins that cause alternative splicing, are found in large numbers in those neurons that degenerate early. For example, the splicing factor PTB is often expressed in nerve cells that produce pathological tau isoforms and degenerate prematurely.

The tau pathology in Alzheimer's disease was long understood as a result of the amyloid plaques. Today, however, the scientific community no longer sees this causal relationship as given, i.e. the protein deposits occur independently of each other and also in healthy people. Nevertheless, there are indications that Aβ plaques in the vicinity of axons can promote the formation of a special form of tau aggregates. However, in contrast to tau fibrils, Aβ peptides are already detectable early in the cortex, particularly in the neocortex and in the entorhinal cortex of the temporal lobe. In the cerebrospinal fluid they may be detected 15–20 years before the onset of dementia. In contrast, the tau pathology correlates directly with neuronal cell death. Never the less, we have to take into account that the blood supply to the brain tissue is reduced in old age which may also explain the observed changes, since vascular and Alzheimer's dementia often occur together.

## Pathological Effects of p-tau and Aβ

As discussed in the first chapter, the vessels of the aging brain are generally less permeable. The changes in endothelial functions and vessel wall thickness (commonly referred to as arteriosclerosis) lead to reduced blood flow and are detectable in many Alzheimer's patients long before the cognitive disorders become apparent. Pathological tau may also play a role by causing constriction of vessels through inhibition of the NO-system (see Fig. 2.4). The capillary networks in brain tissue are normally expanded by nitrogen monoxide (NO), the gaseous neuromodulator released from synapses. The neuronal NO-synthase (nNOS) produces a lot of NO at high neuronal activity, which quickly diffuses into the pericytes around the endothelium of the capillaries. The pericytes then relax, the capillaries dilate and blood flow increases. This phenomenon described as "neurovascular coupling" above is disturbed by high levels of phosphorylated tau that prevents the binding of nNOS to a postsynaptic protein (PSD95) and thus reduces the release of NO.

But we must realize that both phenomena, the formation of senile plaques and pathological tau, cannot be the causal triggers of neurodegeneration in Alzheimer's disease. They occur as a result of yet to be identified molecular changes in neurons, which then lead to Aβ aggregation, phosphorylation of Tau (p-tau), and finally to neuronal cell death. Tau fibrils and pathological Aβ peptides are thus closely correlated with senescence, but not the initial triggers of the disease. Intriguingly, glial cells apparently also play an important pathomechanistic role, because the removal of aging astrocytes and microglia reduces the formation of tau fibrils and limit gliosis, i.e. the proliferation and enlargement of glial cells, in a mouse model of tau-associated diseases.

Structures in the medial temporal lobe, in particular the hippocampus and the entorhinal cortex organizing the in- and output to the hippocampus, are particularly sensitive to p-tau deposits. This is of clinical relevance because the temporal lobe contains phylogenetically old structures that can be damaged by chronically high levels of stress hormones (e.g. by cortisol). Cortisol-dependent effects can also be caused by severe psychological stress. Several links have been described between dementia and severe childhood traumas, but also with depression. Finally, it is known that in post-traumatic stress disorders, the hippocampus and the gyrus cinguli atrophy. Interestingly, affected patients have a higher risk of developing dementia after 5–10 years as compared to people who have not experienced severe psychological trauma before.

A new area of research is focusing on the cortico-hippocampal connections that could underlie stress-induced disorders in the brain. These neural networks are particularly active in early, preclinical stages of dementia. In contrast, their activity decreases when manifest memory disorders are diagnosed. Early hypersensitivity of neuronal circuits is not only recognizable by imaging methods and electroencephalography (EEG), but also by an increased occurrence of epileptic discharge up to real seizures, especially in the early stages of Alzheimer's disease. Patients with changes in the presenilin-1 (PS1) gene are particularly affected. The increased neuronal activity could lead to a faster spread of pathological peptides in the brain. As mentioned above, an excitatory effect on neuronal membranes can be assumed for the Aβ peptides, whereas pathological Tau forms, due to a decrease of the mobility of synaptic vesicles, generally reduce neuronal activity.

**In a Nutshell**
- Alzheimer's typical fibrils, the tangles, consist of highly phosphorylated and aggregated tau protein. Tau is diagnostically important and can be detected by imaging methods (PET) or in the cerebrospinal fluid.
- Increased phosphorylation of tau or modifications of the tau gene (MAPT) disturb the binding of the protein to microtubules and its degradation.
- Phospho-tau inhibits the nitric oxide-induced dilation of capillaries and thus contributes to the reduced perfusion of brain tissue.
- In contrast to amyloid plaques, the tau pathology correlates well with areas of neuronal degeneration and severity of memory problems.
- In the medial temporal lobe areas necessary for learning and memory, pathological tau deposits are found in most Alzheimer's patients.

## Episodic Memory is Already Impaired in the Early Stage of Alzheimer's

Patients with damage to the hippocampus or to the entorhinal cortex have difficulty storing newly acquired information. The retrieval of recently stored information is also impaired. As described in the first chapter, the hippocampus is necessary for long-term storage of the declarative (word and number bound) memory content in the neuronal networks of the neocortex. It is likely that explicit content is stored several times in various cortical areas, if it has been learned under personally and emotionally relevant conditions.

The context of important personal memories is repeatedly called up and newly stored with each activation. The associated content will thus be retrieved into consciousness, that is, into working memory, independently of the hippocampus. This explains why decades-old and emotionally coloured experiences are often well preserved in patients with Alzheimer's dementia, even in the advanced stage of the disease.

Procedural (mainly motor) memory contents, for example the ability to swim or to dance, are not primarily affected in Alzheimer's disease. Early learned movement programs that are stored in the subcortical basal ganglia and in the intact motor cortex remain unaffected for a long time. In the late phase of the disease, however, the Tau fibrils are detectable not only in the archi- and neocortex, but also in the basal nuclei as well as in the diencephalon and mesencephalon, followed by neuronal degeneration in these areas. Intriguingly, the amyloid plaques initially do not occur in the archicortex, but rather in the neocortex and are only detectable much later in deeper lying brain regions. This difference also speaks against a causal connection between plaques, neuronal degeneration and clinical symptoms.

In search of neuroanatomical structures that are involved in explicit memory in addition to the cortex, researchers took a closer look at the brainstem a few years ago. Here, in the pons, near the fourth ventricle, lies a small but well-defined blue-appearing area, the locus coeruleus, which was already mentioned in the context of Parkinson's disease. This most important noradrenergic nucleus is about 15 mm in size and contains the dark neuromelanin pigment, which makes it recognizable to the naked eye. This nucleus supplies almost the entire brain via a far-flung network of axons with noradrenaline, the transmitter produced from dopamine via dopamine β-hydroxylase (Fig. 2.5).

Noradrenaline contributes as a neuromodulator to attention and motivation. A number of animal experiments have shown that it also supports cellular remodeling that allows for long-term storage of new memory content. Interestingly, not only are tau deposits detectable in the locus coeruleus in old age, but also in children. Some experts assume that pathological tau detected in affected noradrenergic neurons reaches the temporal lobe and later the neocortex, where it may be released trans-synaptically and taken up by nerve cells (in a prion-like manner as discussed in the next paragraph).

The justification for this hypothesis is based on memory tests that were combined with imaging studies. In doing so, younger subjects were able to solve the tasks set much better than older people. The locus coeruleus of the older participants with good memory was morphologically similar to the locus coeruleus of younger subjects, i.e. it contained fewer age-related changes that could indicate an early form of degeneration. Analogous to Parkinson's disease, early changes should therefore also be sought in the brainstem of suspected Alzheimer's patients.

In the following, I would like to discuss some of the cellular mechanisms that could transmit Alzheimer's pathology and thus neuronal degeneration. In particular, the processes relevant for the intra- and extracellular transport of Aβ peptides and tau are of major interest.

## 2.3.5   The Prion Theory in Alzheimer's Disease

As described for α-synuclein in Parkinson's disease, Tau protein can be passed from one neuron to another. In the healthy recipient cell, the formation of aggregates would then be promoted in a prion-like way. This particularly affects neurons that are connected to each other via synaptic contacts. In the case of pyramidal neurons with large dendritic trees in the cortex or hippocampus, the Tau aggregates first occur in the distal, more distant dendrites, then in the cell body and only late in the axon, which speaks for a transfer of pathological Tau forms from the input contact point (the pre-synapse) to the output contact point (post-synapse).

This hypothesis could also explain the neuropathological staging system defined by Heiko Braak relevant for most patients with Alzheimer's disease: Tau pathology spreads from the lower, occipital-temporal gyri of the temporal lobe (stage I) into the parahippocampal gyrus (with the entorhinal cortex in front of it) on the inside of the temporal lobe (stage II), from there further into the hippocampus (stage III), into the adjacent temporal neocortex (stage IV), and finally into secondary (stage V) and primary cortical areas (stage VI). However, recent imaging studies have shown that in individual patients, tau pathology with only minor involvement of the medial temporal lobes can also be found.

In addition to neurons, glial cells take up extracellular aggregates, too. The functional significance of tau accumulation in glial cells was only recently demonstrated: Pathological tau forms (3R-tau) are found in large numbers in astrocytes of Alzheimer's patients, and overexpression of tau in hippocampal astroglia of mice results in a disturbance of mitochondrial mobility, reduced neurogenesis, and impaired spatial memory.

In the case of the transferability of Tau, we must assume a pathology limited to the brain, because—in contrast to Parkinson's disease—the deposits typical of Alzheimer's disease are not found in the peripheral or enteric nervous system of humans (only very little β-amyloid and APP were demonstrated in the intestine of Alzheimer's patients, but also in healthy people). However, in animal models of Alzheimer's disease, for example in the intestine of APP-overexpressing mice, β-amyloid aggregates and inflammatory reactions in the enteric ganglia as well as changes in the microbiome can clearly be detected. Even so, the locus coeruleus affected early in Alzheimer's disease has no direct anatomical connection to the peripheral or enteric nervous system. Hence, it remains an open question if an early diagnosis of Alzheimer's disease based on a peripheral tissue sample in Alzheimer's will be possible in the future.

How do Aβ peptides or pathological Tau forms travel from one nerve cell to another? As with synuclein, the proteins are probably released via exosomes (small membrane-enclosed vesicles, Fig. 2.3). This is supported by the fact that exosomal membrane proteins have been found in amyloid plaques. On the other hand, microscopically small, tubular connections between the cells (so-called nano-membrane tubes) have been observed and could also allow the exchange of peptides and smaller proteins between cells.

It must be emphasized that, in contrast to genuine prion diseases, there is currently no evidence for transmission of Alzheimer-typical tau fibrils from human to human or from animal to human. It has indeed been reported that new Aβ aggregates are formed in patients after transplantation of the dura mater or after administration of animal-derived growth hormone. These preparations regularly contained soluble tau, but the decisive pathological tau forms, e.g. fibrils, were not found in the brain of patients who had received foreign tissue or exogenous hormone preparations.

Alternatively to exosomal transport, tau molecules could be released directly into the extracellular space. However, these are mainly fragments of the complete tau protein that can no longer bind to microtubules or form aggregates. Apparently, only intact and exocytosed tau is taken up by adjacent nerve cells, probably mediated by a member of the LDL-receptor-family, the low-density lipoprotein receptor-related protein 1 (LRP1).

LRP1 also interacts with the lipid-binding ApoE, which is primarily produced by astrocytes. ApoE is a polymorphic gene, i.e. it comes in 3 main forms (E2, E3 and E4). Those with two ApoE4 alleles have a higher chance of developing Alzheimer's disease than those with only one allele (almost all homozygous carriers of ApoE4 are affected at the age of 80 or older). The E2 variant, on the other hand, is interesting in that it delays the onset of Alzheimer's dementia.

It is assumed that the two ApoE forms have different effects on the phagocytotic activity of astrocytes. While ApoE2 makes it easier for astrocytes to eliminate defective synapses in their vicinity, ApoE4 is said to have the opposite effect, so that synapses can be attacked more easily by the complement system. However, these observations have not yet been confirmed.

Neurons take up the ApoE protein via their membrane-bound ApoE transporters. Extracellular Tau or Tau oligomers apparently bind to LRP1 or ApoE transporters, but only LRP1 causes uptake into the cell. The discovery of Tau-binding LDL receptors underscores the importance of lipid metabolism in the disease. Interestingly, among the Alzheimer's genes are not only those involved in the production and transport of proteins, but also of triglycerides and cholesterol (e.g. ApoE4). Cholesterol esters, which are formed at high cholesterol levels in the blood, lead to increased amyloid formation and Tau deposition. They probably also interfere with protein degradation in the proteasome.

**In a Nutshell**

- Explicit (declarative), but not implicit memory content is primarily lost in Alzheimer's disease. The hippocampus within the temporal lobe is particularly affected.
- The locus coeruleus, the largest noradrenaline-producing nucleus in the brainstem, early shows pathological Tau fibrils that can be passed on to the cortex in a prion-like manner.
- Glial cells (astrocytes) also take up pathological Tau proteins. These disturb the mitochondrial mobility in animal models of Alzheimer's disease.
- Intact Tau released by exocytosis is taken up by neurons via a protein of lipid metabolism (LRP1), which interacts with ApoE, another lipid-binding protein and genetic risk factor.

## The Relevance of Cerebrospinal Fluid Drainage for Neurodegeneration

At the end of this section, an alternative theory of the spread of typical Alzheimer's pathology should be presented. It may be that the intercellular transfer of Aβ peptides, phosphorylated tau, or α-synuclein in Parkinson's plays only a minor role in the pathogenesis of both diseases. Alternatively, it would be assumed that pathological proteins simply accumulate extracellularly over the long term, as the flow of cerebrospinal fluid in the glymphatic system is reduced at older age (see Fig. 2.4).

This hypothesis is based on the observation that a contrast agent injected into the cerebrospinal fluid is distributed in a way that exactly corresponds to the stages of the spread of deposits proposed by Heiko Braak: The pathology begins on the mediobasal side of the frontal lobe and in the gyrus cinguli, then follows the course of the large cerebral arteries into the limbic system and posterior cranial fossa, and finally ends in the neocortex. The transport of fluid is caused by the pulsation of the cerebral arteries and clearly resembles the distribution of Alzheimer's associated protein aggregates, which was confirmed in elaborate magnetic resonance imaging studies in a large number of patients.

Similarly to a disturbed lymph flow in our extremities, the disturbed flow of cerebrospinal fluid in old age (and also in disturbed sleep) could therefore lead to an increased concentration of pathological proteins in the extracellular space. During the night, the drainage of cerebrospinal fluid and thus the transport of proteins is stimulated especially in deep non-REM sleep. However, these sleep phases hardly occur anymore in people over the age of 60. The same physicochemical criteria that play a role in the formation of the intracellular condensates discussed above (Fig. 2.11) would then lead to the formation of the extracellular protein aggregates. This hypothesis also explains the spatial and temporal course of the disease in most patients.

However, it is possible that both phenomena are involved in the pathogenesis, i.e. the intercellular transport of pathological peptides and proteins and a reduced flow of cerebrospinal fluid could jointly promote the pathology of Alzheimer's or Parkinson's disease. In the latter, characteristic sleep disorders are found long before the onset of symptoms, in particular, during the REM sleep phases characterized by rapid eye movements and increased blood pressure.

## 2.4  Inflammatory Components of Alzheimer's and Parkinson's Disease

It has been mentioned several times that inflammatory processes are of great importance in the context of neurodegenerative diseases. Monocytes, the precursors of macrophages, as well as CD4- and CD8-positive T-cells migrate from the blood into the brain. But also the glia, in particular microglia and astrocytes, play an important role in disease pathogenesis. Interestingly, several genes that have been identified as risk factors for neuronal degeneration are expressed in microglial cells which, as described above, can trigger an inflammatory reaction in the brain.

The occasional observation that Alzheimer's and Parkinson's disease occur less frequently in older people who take anti-inflammatory drugs may also be relevant in this context. However, it is controversially discussed whether long-term pain medications, such as non-steroidal anti-rheumatics, may have a favorable effect on the onset or course of neurodegenerative diseases, because the statistical effects are rather low in the studies that have been carried out so far.

Nevertheless, it is now well established that frequent infections in old age accelerate the cognitive decline. This negative effect on the development of dementia is probably caused by an increase in cytokines, e.g. in the tumor necrosis factor, TNFα. An important source of TNFα in the brain is microglia with their macrophage-like properties. They phagocytize whole cells, cell components, but also synaptic contacts or the α-synuclein or amyloid aggregates and fight invading pathogens by releasing inflammatory mediators and cytokines.

In addition, Aβ oligomers are able to bind to specific microglial receptors (CD36, TLR4/6, RAGE, TREM2) and activate them. The reactions of microglia to Aβ peptides may even be protective, as people with mutations in the TREM2 gene have an increased risk of Alzheimer's disease. A reduction in these clearly positive properties of microglia and macrophages is generally observed as part of the aging process (senescence), and apparently due to an increased level of prostaglandin E2 (PGE2) which acts as immunosuppressant in the brain.

The ambivalent character of microglia is particularly evident when over-activated. The microglia cells then begin to divide and take on an amoeboid form (instead of their normally more branching morphology). Intracellularly, they form special protein complexes, the inflammasomes. which are part of the innate immune response and function as specific sensors that can quickly respond to foreign proteins and inflammatory signals. Caspase-1 then forms the active inflammatory mediators (in particular IL-1β and TNFα) from precursors of the interleukins. After release of these cytokines into the extracellular space, specific neuronal signaling pathways are activated via the corresponding receptors, including the p38-MAP kinase signal pathway, which can cause strong phosphorylation of tau proteins and thus further damage.

Many patients suffering from Alzheimer's disease therefore show microglial activation specifically in regions of high p-tau concentrations. In addition to the synuclein pathology described above, an inflammatory component can also be detected in the cortex of Parkinson's patients, especially when dementia is already present. This inflammatory reaction characterized by activated microglia, infiltration of T lymphocytes and pro-inflammatory cytokines is significantly less pronounced in patients without dementia.

Ultimately, inflammatory cells lead to an accelerated loss of healthy synapses and the death of nerve cells. The complement system, especially the factors C1q and C3b, play an important role in this process. Activated microglia produces more of the C1 complex, which can bind to synaptic membranes. This leads to classical complement activation and subsequent phagocytosis (clearance function of microglia). Antibodies that block this process have been shown to protect synapses from being removed by microglia.

On the other hand, it must also be noted that, at least in animal models of Alzheimer's disease, neuronal degeneration can also be halted without any changes to the surrounding inflammation, for example by turning off the tau-binding vesicle protein synaptogyrin-3 in neurons. Furthermore, microglial cells take up extracellular aggregates, such as β-amyloid, at the beginning of the disease. Finally, damaged cells and synapses are cleared away by microglia. Defective synapses disturb the entire cell by allowing calcium influx and impairment of mitochondria, so that microglial cells are also responsible for the maintenance of neuronal networks. The microglia in the brain is therefore to be seen as a double-edged sword.

**In a Nutshell**

- Frequent infections in old age accelerate cognitive decline. The tumor necrosis factor TNFα plays an important role in this.
- Aβ-oligomers bind to special microglia receptors (TREM2). People with mutations in the TREM2 gene have an increased risk of developing Alzheimer's dementia.
- Microglia activation occurs particularly in regions with pronounced Tau pathology.
- In the early stages of a neurodegenerative disease, microglia may also act protectively by taking up and disposing of defective synapses and protein aggregates (e.g. β-amyloid).

## 2.5    Viral Infections in Neurodegenerative Diseases

In relation to neurodegenerative diseases, various reports and animal experiments suggest that viruses can particularly affect dopaminergic neurons in the midbrain. An increased incidence of Parkinson's syndromes, but also of Alzheimer's-typical changes, was already observed during the Spanish flu (1918–1920) and following the bird flu epidemics caused by the H5N1 virus. It is assumed that viral infections trigger the chronic activation of microglia, which ultimately leads to neurodegeneration via the inflammatory changes described above but not direct damage of nerve cells by viruses constantly present in the brain. Parkinsonism has also been observed after infection with coxsackie, encephalitis or human immunodeficiency viruses.

As part of the worldwide Covid pandemic the possible involvement of corona viruses in neuronal cell death has come into the focus of many laboratories worldwide. In December 2019, a new coronavirus (SARS-CoV-2, Fig. 2.19) appeared in Wuhan, China, which, in addition to a number of proteins, contains a single-stranded RNA molecule. After multiplication in the host cells the virus produces an acute respiratory syndrome in the lungs (SARS stands for **S**evere **A**cute **R**espiratory **S**yndrome).

The mortality of coronavirus disease (Covid)-19 is 1–3% across all age groups. Of the elderly and people with pre-existing conditions (e.g. high blood pressure, diabetes, lung and heart diseases), up to 20% of those infected die. By April 2021, there were more than 150 million infected people and 3.2 million victims. In addition, 5–10% of those infected suffer from long-Covid syndrome, a chronic fatigue with sometimes severe effects on the functioning of our brain. This infection is the world's worst health crisis since the second World War. Unfortunately, we also had to experience the enormous impact the disease has on our health and economic systems.

The membrane receptor ACE2 (**A**ngiotensin-**C**onverting **E**nzyme receptor) required for the entry of the SARS-CoV-2 virus into target cells is found in the respiratory tract, but also in nerve and glial cells as well as on the inner lining of blood vessels, the endothelium. The large, mushroom-shaped spike protein (Fig. 2.19) allows the virus to fuse with the target cell's lipid membrane, so that the viral RNA enters the cell.

In addition to the epithelial cells in the nose and mouth, the pneumocytes in the lungs and the endothelial cells of blood vessels are infected. This may lead to lung damage and to circulatory disorders, which can end in multi-organ failure. If there is a high virus load in the blood (viremia),

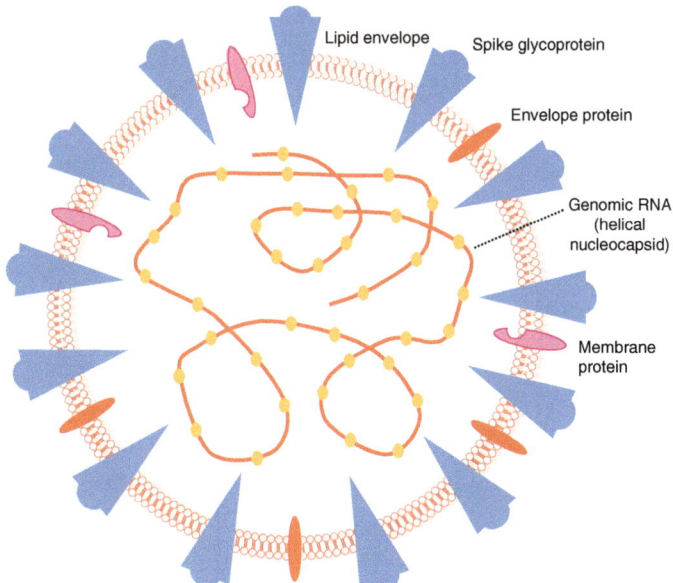

**Fig. 2.19**   The RNA of the SARS-CoV-2 virus contains the information for all virus proteins, in particular for the Spike, envelope, membrane and nucleocapsid proteins. Other proteins are essential for replication. After infection and uptake of the virus into the cytoplasm, the ribosomes of the host cell take over the production of these approximately 30 proteins. New virus particles are then assembled in the Golgi apparatus and secreted to infect other cells

the virus enters the brain through the endothelium of the brain vessels and also through the first cranial nerve, the olfactory nerve. From there, the virus spreads further into the olfactory bulb at the bottom of the frontal lobe. The olfactory support cells in the epithelium, but not the olfactory receptor cells, in our upper nasal cavity show a particularly high infection rate with SARS-CoV-2.

Neurological symptoms are found in up to 80% of people infected. In general, they are to be expected as non-specific accompanying symptoms of severe infection, sepsis with high fever or due to oxygen deficiency. In addition to head- and muscle -aches, fatigue, smell and taste disorders in around 90% of patients, a severe inflammation of the brain (encephalitis) or a stroke may occur. In histopathological examinations, a microglial inflammatory reaction and changes in the small blood vessels were described. The wall of the affected vessels becomes thinner and starts to leak, so that circulatory problems follow. In addition, peripheral neuropathies and myopathies as well as the feared Guillain-Barre syndrome with paralysis of the muscles are observed.

However, SARS-CoV-2 virus is rarely detected in large amounts in nervous tissue or in the cerebrospinal fluid. Although the virus can in principle penetrate into all ACE2-positive neurons, apparently the immune reaction started by the virus in the brain is primarily responsible for the symptoms described. The innate immune system first recognizes viral RNA in the respiratory tract and triggers the expression of interferons. As a result, cytotoxic T cells (CD8-positive) and helper T cells (CD4-positive) are activated. The latter in turn stimulate B-lymphocytes, which form antibodies against virus proteins and thus block the further uptake of viruses into target cells.

As part of these immunological processes, however, some interferons and cytokines may be released in such large numbers that an excessive immune reaction, a so-called cytokine storm, results. This explains why in some patients a steroid (cortisol) treatment that suppresses the immune system results in clinical improvements, even though this therapy is usually contraindicated in viral infections. In the brain, exhausted T cells are found in the months after an infection. They may be involved in neurological long Covid, since they no longer have a complete repertoire of immunologically relevant proteins available. Interestginly, the precursors of macrophages, the monocytes, also reveal a reduced cellular metabolism.

It remains to be seen to what extent these changes in Covid-19 patients lead to neuronal degeneration in the long term. Intriguingly, people at increased risk for Alzheimer's disease (e.g. ApoE4 carriers) are more likely to be infected than others, and in one third of demented patients, a Covid infection accelerates the loss of cognitive abilities. The neurological symptoms also worsen in Parkinson's patients who become infected with SARS-CoV-2. They complain of increased tremor, muscle stiffness, fatigue, depression and pain during the course of Covid.

The future will show whether the SARS virus can directly attack the substantia nigra and whether Parkinson's syndromes will then occur more frequently. So far, there are only few descriptions that point in this direction. However, long-term neurological complications are certainly to be expected in many Covid-19 patients, including children and adolescents. The symptoms include, in particular, mood swings, concentration and memory disorders, fatigue or dizziness. The large number of chronically ill people prompted the German Neurological Society to issue its own treatment guidelines.

In addition to the usual symptomatic therapies and known anti-viral drugs, research is currently being conducted on therapeutic antibodies and peptides that can block the virus from entering nerve cells. An interesting approach is the inhibition of binding of the virus to $\alpha5\beta1$-integrin in the neuronal plasma membrane, since the spike protein of SARS-CoV-2 can not only bind to ACE2, but also to integrins via an RGD motif, which consists of the three amino acids arginine, glycine and asparagine (Arg, R; Gly, G; Asp, D).

## In a Nutshell

- Parkinson's syndrome and Alzheimer-typical symptoms can occur as a result of viral infections, possibly also as a consequence of Covid-19.
- The new infection with SARS-CoV-2 viruses from the Corona family that occurred at the end of 2019 often results in muscle ache, fatigue and smell deficiency.
- The virus enters the brain via the endothelium of the brain vessels or via the first cranial (olfactory) nerve.
- Up to 10% of those infected suffer from a chronic fatigue syndrome, which is primarily attributed to an ongoing microglial inflammatory reaction in the brain.

# Further Reading

# Aging and Neuronal Degeneration

Andreone BJ, Larhammar M, Lewcock JW (2019) Cell death and neurodegeneration. Cold Spring Harb Perspect Biol 12:a036434

Bartels T, De Schepper S, Hong S (2020) Microglia modulate neurodegeneration in Alzheimer's and Parkinson's diseases. Science 370:66–69

Berson A, Nativio R, Berger SL, Bonini NM (2018) Epigenetic regulation in neurodegenerative diseases. Trends Neurosci 41:587–598

Bonnar O, Hall CN (2020) First, tau causes NO problem. Nat Neurosci 23:1035–1036

Braak H, Del Tredici K (2016) Potential pathways of abnormal Tau and α-Synuclein dissemination in sporadic Alzheimer's and Parkinson's diseases. Cold Spring Harb Perspect Biol 8:a023630

Chen Y, Qin C, Huang J, Tang X, Liu C, Huang K, Xu J, Guo G, Tong A, Zhou L (2020) The role of astrocytes in oxidative stress of central nervous system: a mixed blessing. Cell Prolif 53:e12781

Cioni J-M, Lin JQ, Holtermann AV, Koppers M, Jakobs MAH, Azizi A, Turner-Bridger B, Shigeoka T, Franze K, Harris WA, Holt CE (2019) Late endosomes act as mRNA translation platforms and sustain mitochondria in axons. Cell 176:56–72

Clarke LE, Liddelow SA, Chakraborty C, Munch AE, Heiman M, Barres BA (2018) Normal aging induces A1-like astrocyte reactivity. Proc Natl Acad Sci USA 115:E1896–E1905

Crispi S, Filosa S (2021) Novel perspectives for neurodegeneration prevention: effects of bioactive polyphenols. Neural Regen Res 16:1411–1412

Cullen NC, Leuzy A, Palmqvist S, Janelidze S et al (2021) Individualized prognosis of cognitive decline and dementia in mild cognitive impairment based on plasma biomarker combinations. Nature Aging 1:114–123

Dahl MJ, Mather M, Düzel S, Bodammer NC, Lindenberger U, Kühn S, Werkle-Bergner M (2019) Rostral locus coeruleus integrity is associated with better memory performance in older adults. Nat Hum Behav 3:1203–1214

Dawson TM, Dawson VL (2017) Mitochondrial mechanisms of neuronal cell death: potential therapeutics. Annu Rev Pharmacol Toxicol 57:437–454

Dawson TM, Golde TE, Lagier-Tourenne C (2018) Animal models of neurodegenerative diseases. Nat Neurosci 21:1370–1379

Delpech JC, Herron S, Botros MB, Ikezu T (2019) Neuroimmune crosstalk through extracellular vesicles in health and disease. Trends Neurosci 42:361–372

Eacker SM, Dawson TM, Dawson VL (2009) Understanding microRNAs in neurodegeneration. Nat Rev Neurosci 10:837–841

Fricker M, Tolkovsky AM, Borutaite V, Coleman M, Brown GC (2018) Neuronal cell death. Physiol Rev 98:813–880

Fujikake N, Shin M, Shimizu S (2018) Association between autophagy and neurodegenerative diseases. Front Neurosci 12:255–255

Gan L, Cookson MR, Petrucelli L, La Spada AR (2018) Converging pathways in neurodegeneration, from genetics to mechanisms. Nat Neurosci 21:1300–1309

Greenhalgh AD, David S, Bennett FC (2020) Immune cell regulation of glia during CNS injury and disease. Nat Rev Neurosci 21:139–152

Guttenplan KA, Liddelow SA (2019) Astrocytes and microglia: models and tools. J Exp Med 216:71–83

Heneka MT, McManus RM, Latz E (2018) Inflammasome signalling in brain function and neurodegenerative disease. Nat Rev Neurosci 19:610–621

Hickman S, Izzy S, Sen P, Morsett L, El Khoury J (2018) Microglia in neurodegeneration. Nat Neurosci 21:1359–1369

Hollville E, Romero SE, Deshmukh M (2019) Apoptotic cell death regulation in neurons. FEBS J 286:3276–3298

Hwang JY, Aromolaran KA, Zukin RS (2017) The emerging field of epigenetics in neurodegeneration and neuroprotection. Nat Rev Neurosci 18:347–361

Jha MK, Kim JH, Song GJ, Lee WH, Lee IK, Lee HW, An SSA, Kim S, Suk K (2018) Functional dissection of astrocyte-secreted proteins: implications in brain health and diseases. Prog Neurobiol 162:37–69

Jucker M, Walker LC (2018) Propagation and spread of pathogenic protein assemblies in neurodegenerative diseases. Nat Neurosci 21:1341–1349

Kam TI, Hinkle JT, Dawson TM, Dawson VL (2020) Microglia and astrocyte dysfunction in Parkinson's disease. Neurobiol Dis 144:105028

Kiral FR, Kohrs FE, Jin EJ, Hiesinger PR (2018) Rab GTPases and membrane trafficking in neurodegeneration. Curr Biol 28:R471–R486

Klimaschewski L, Claus P (2021) Fibroblast growth factor signalling in the diseased nervous system. Mol Neurobiol 58:3884–3902

Kulkarni A, Chen J, Maday S (2018) Neuronal autophagy and intercellular regulation of homeostasis in the brain. Curr Opin Neurobiol 51:29–36

Leeman DS, Hebestreit K, Ruetz T, Webb AE et al (2018) Lysosome activation clears aggregates and enhances quiescent neural stem cell activation during aging. Science 359:1277–1283

Li Q, Haney MS (2020) The role of glia in protein aggregation. Neurobiol Dis 143:105015

Liddelow SA, Barres BA (2017) Reactive astrocytes: production, function, and therapeutic potential. Immunity 46:957–967

Liddelow SA, Sofroniew MV (2019) Astrocytes usurp neurons as a disease focus. Nat Neurosci 22:512–513

Lim YJ, Lee SJ (2017) Are exosomes the vehicle for protein aggregate propagation in neurodegenerative diseases? Acta Neuropathol Commun 5:64

Ma S, Sun S, Geng L, Song M, Wang W et al (2020) Caloric restriction reprograms the single-cell transcriptional landscape of rattus norvegicus aging. Cell 180:984–1001

Mathieu C, Pappu RV, Taylor JP (2020) Beyond aggregation: pathological phase transitions in neurodegenerative disease. Science 370:56–60

Mattugini N, Bocchi R, Scheuss V, Russo GL, Torper O, Lao CL, Götz M (2019) Inducing different neuronal subtypes from astrocytes in the injured mouse cerebral cortex. Neuron 103:1086–1095

McKenzie BA, Dixit VM, Power C (2020) Fiery cell death: pyroptosis in the central nervous system. Trends Neurosci 43:55–73

Nedergaard M, Goldman SA (2020) Glymphatic failure as a final common pathway to dementia. Science 370:50–56

Neefjes J, van der Kant R (2014) Stuck in traffic: an emerging theme in diseases of the nervous system. Trends Neurosci 37:66–76

Neukomm LJ, Burdett TC, Seeds AM, Hampel S et al (2017) Axon death pathways converge on Axundead to promote functional and structural axon disassembly. Neuron 95:78–91

Ochoa Thomas E, Zuniga G, Sun W, Frost B (2020) Awakening the dark side: retrotransposon activation in neurodegenerative disorders. Curr Opin Neurobiol 61:65–72

Pan C, Locasale JW (2021) Targeting metabolism to influence aging. Science 371:234–235

Peng C, Trojanowski JQ, Lee VMY (2020) Protein transmission in neurodegenerative disease. Nat Rev Neurol 16:199–212

Qian H, Kang X, Hu J, Zhang D, Liang Z et al (2020) Reversing a model of Parkinson's disease with in situ converted nigral neurons. Nature 582:550–556

Riera CE, Dillin A (2015) Can aging be "drugged"? Nat Med 21:1400–1405

Saez-Atienzar S, Masliah E (2020) Cellular senescence and Alzheimer disease: the egg and the chicken scenario. Nat Rev Neurosci 21:433–444

Salter MW, Stevens B (2017) Microglia emerge as central players in brain disease. Nat Med 23:1018–1027

Sandsmark DK, Bashir A, Wellington CL, Diaz-Arrastia R (2019) Cerebral microvascular injury: a potentially treatable endophenotype of traumatic brain injury-induced neurodegeneration. Neuron 103:367–379

Sweeney MD, Kisler K, Montagne A, Toga AW, Zlokovic BV (2018) The role of brain vasculature in neurodegenerative disorders. Nat Neurosci 21:1318–1331

Vainchtein ID, Molofsky AV (2020) Astrocytes and microglia: in sickness and in health. Trends Neurosci 43:144–154

Wang C, Telpoukhovskaia MA, Bahr BA, Chen X, Gan L (2018) Endo-lysosomal dysfunction: a converging mechanism in neurodegenerative diseases. Curr Opin Neurobiol 48:52–58

Wang C, Yue H, Hu Z, Shen Y, Ma J, Li J, Wang XD, Wang L, Sun B, Shi P, Wang L, Gu Y (2020) Microglia mediate forgetting via complement-dependent synaptic elimination. Science 367:688–694

Wertz MH, Mitchem MR, Pineda SS, Hachigian LJ et al (2020) Genome-wide in vivo CNS screening identifies genes that modify CNS neuronal survival and mHTT toxicity. Neuron 106:76–89

Yerbury JJ, Farrawell NE, McAlary L (2020) Proteome homeostasis dysfunction: a unifying principle in ALS pathogenesis. Trends Neurosci 43:274–284

# Parkinson's Disease

Armstrong MJ, Okun MS (2020) Diagnosis and treatment of Parkinson disease: a review. JAMA 323:548–560

Bartels T, De Schepper S, Hong S (2020) Microglia modulate neurodegeneration in Alzheimer's and Parkinson's diseases. Science 370:66–69

Cammisuli D, Ceravolo R, Bonuccelli U (2020) Non-pharmacological interventions for Parkinson's disease mild cognitive impairment: future directions for research. Neural Regen Res 15:1650–1651

Cao K, Tait SWG (2019) Parkin inhibits necroptosis to prevent cancer. Nat Cell Biol 21:915–916

Carling PJ, Mortiboys H, Green C, Mihaylov S et al (2020) Deep phenotyping of peripheral tissue facilitates mechanistic disease stratification in sporadic Parkinson's disease. Prog Neurobiol 187:101772

Diederich NJ, Uchihara T, Grillner S, Goetz CG (2020) The evolution-driven signature of Parkinson's disease. Trends Neurosci 43:475–492

Fares MB, Jagannath S, Lashuel HA (2021) Reverse engineering Lewy bodies: how far have we come and how far can we go? Nat Rev Neurosci 22:111–131

Fellner L, Irschick R, Schanda K, Reindl M, Klimaschewski L, Poewe W, Wenning GK, Stefanova N (2013) Toll-like receptor 4 is required for α-synuclein dependent activation of microglia and astroglia. Glia 61:349–360

Hunn BHM, Cragg SJ, Bolam JP, Spillantini MG, Wade-Martins R (2015) Impaired intracellular trafficking defines early Parkinson's disease. Trends Neurosci 38:178–188

Kalia LV, Lang AE (2015) Parkinson's disease. Lancet 386:896–912

Kam TI, Hinkle JT, Dawson TM, Dawson VL (2020) Microglia and astrocyte dysfunction in Parkinson's disease. Neurobiol Dis 144:105028

Kouli A, Camacho M, Allinson K, Williams-Gray CH (2020) Neuroinflammation and protein pathology in Parkinson's disease dementia. Acta Neuropathol Commun 8:211

Navarro-Romero A, Montpeyó M, Martinez-Vicente M (2020) The emerging role of the lysosome in Parkinson's disease. Cell 9:2399

Poewe W, Seppi K, Tanner CM, Halliday GM, Brundin P, Volkmann J, Schrag AE, Lang AE (2017) Parkinson disease. Nature Rev Dis Primers 3:17013

Qian H, Kang X, Hu J, Zhang D, Liang Z et al (2020) Reversing a model of Parkinson's disease with in situ converted nigral neurons. Nature 582:550–556

Sorrentino ZA, Giasson BI (2019) Exploring the peripheral initiation of Parkinson's disease in animal models. Neuron 103:547–549

Surmeier DJ, Obeso JA, Halliday GM (2017) Selective neuronal vulnerability in Parkinson disease. Nat Rev Neurosci 18:101–113

Wang Z, Becker K, Donadio V, Siedlak S, Yuan J et al (2020) Skin α-Synuclein aggregation seeding activity as a novel biomarker for Parkinson disease. JAMA Neurol 78:1–11

Zeng XS, Geng WS, Jia JJ, Chen L, Zhang PP (2018) Cellular and molecular basis of neurodegeneration in Parkinson disease. Front Aging Neurosci 10:109

# Dementia and Alzheimer's Disease

Bartels T, De Schepper S, Hong S (2020) Microglia modulate neurodegeneration in Alzheimer's and Parkinson's diseases. Science 370:66–69

Bellenguez C, Grenier-Boley B, Lambert JC (2020) Genetics of Alzheimer's disease: where we are, and where we are going. Curr Opin Neurobiol 61:40–48

Braak H, Del Tredici K (2016) Potential pathways of abnormal Tau and α-Synuclein dissemination in sporadic Alzheimer's and Parkinson's diseases. Cold Spring
Harb Perspect Biol 8:a023630

Chang CW, Shao E, Mucke L (2021) Tau: enabler of diverse brain disorders and
target of rapidly evolving therapeutic strategies. Science 371:eabb8255

Chen X, Gan L (2019) An exercise-induced messenger boosts memory in
Alzheimer's disease. Nat Med 25:20–21

Darling AL, Shorter J (2020) Atomic structures of amyloid-β oligomers illuminate
a neurotoxic mechanism. Trends Neurosci 43:740–743

Edwards FA (2019) A unifying hypothesis for Alzheimer's disease: from plaques to
neurodegeneration. Trends Neurosci 42:310–322

Harris SS, Wolf F, De Strooper B, Busche MA (2020) Tipping the scales: pep-
tide-dependent dysregulation of neural circuit dynamics in Alzheimer's disease.
Neuron 107:417–435

Klimmt J, Dannert A, Paquet D (2020) Neurodegeneration in a dish: advancing
human stem-cell-based models of Alzheimer's disease. Curr Opin Neurobiol
61:96–104

Kumar DKV, Choi SH, Washicosky KJ, Eimer WA, Tucker S et al (2016)
Amyloid-β peptide protects against microbial infection in mouse and worm
models of Alzheimer's disease. Sci Transl Med 8:340–372

Lemche E (2018) Early life stress and epigenetics in late-onset Alzheimer's dementia: a systematic review. Curr Genom 19:522–602

Minhas PS, Latif-Hernandez A, McReynolds MR, Durairaj AS et al (2021)
Restoring metabolism of myeloid cells reverses cognitive decline in ageing.
Nature 590:122–128

Moreno-Jiménez EP, Flor-García M, Terreros-Roncal J, Rábano A et al (2019)
Adult hippocampal neurogenesis is abundant in neurologically healthy subjects
and drops sharply in patients with Alzheimer's disease. Nat Med 25:554–560

Nedergaard M, Goldman SA (2020) Glymphatic failure as a final common pathway
to dementia. Science 370:50–56

Rexach J, Geschwind D (2020) Selective neuronal vulnerability in Alzheimer's disease: a modern holy grail. Neuron 107:763–765

Saez-Atienzar S, Masliah E (2020) Cellular senescence and Alzheimer disease: the
egg and the chicken scenario. Nat Rev Neurosci 21:433–444

Sierksma A, Escott-Price V, De Strooper B (2020) Translating genetic risk of
Alzheimer's disease into mechanistic insight and drug targets. Science 370:61–66

Small SA, Petsko GA (2015) Retromer in Alzheimer disease, Parkinson disease and
other neurological disorders. Nat Rev Neurosci 16:126–132

Small SA, Simoes-Spassov S, Mayeux R, Petsko GA (2017) Endosomal traffic jams
represent a pathogenic hub and therapeutic target in Alzheimer's disease. Trends
Neurosci 40:592–602

Trambauer J, Fukumori A, Steiner H (2020) Pathogenic Aβ generation in familial Alzheimer's disease: novel mechanistic insights and therapeutic implications. Curr Opin Neurobiol 61:73–81

van der Kant R, Goldstein LSB, Ossenkoppele R (2020) Amyloid-β-independent regulators of tau pathology in Alzheimer disease. Nat Rev Neurosci 21:21–35

Walsh DM, Selkoe DJ (2020) Amyloid β-protein and beyond: the path forward in Alzheimer's disease. Curr Opin Neurobiol 61:116–124

# Covid-19

Amruta N, Chastain WH, Paz M, Solch RJ, Murray-Brown IC, Befeler JB, Gressett TE, Longo MT, Engler-Chiurazzi EB, Bix G (2021) SARS-CoV-2 mediated neuroinflammation and the impact of COVID-19 in neurological disorders. Cytokine Growth Factor Rev 58:1–15

Cataldi M, Pignataro G, Taglialatela M (2020) Neurobiology of coronaviruses: potential relevance for COVID-19. Neurobiol Dis 143:105007

Finsterer J (2020) Putative mechanisms explaining neuro-COVID. J Neuroimmunol 350:577453

Gatto EM, Fernandez Boccazzi J (2020) COVID-19 and neurodegeneration: what can we learn from the past? Eur J Neurol 27:e45–e45

Hosp JA, Dressing A, Blazhenets G, Bormann T, Rau A et al (2021) Cognitive impairment and altered cerebral glucose metabolism in the subacute stage of COVID-19. Brain 144:1263–1276

Hu B, Guo H, Zhou P, Shi ZL (2021) Characteristics of SARS-CoV-2 and COVID-19. Nat Rev Microbiol 19:141–154

Lee MH, Perl DP, Nair G, Li W et al (2020) Microvascular injury in the brains of patients with Covid-19. N Engl J Med 384:481–483

Lempriere S (2020) SARS-CoV-2 and the brain to be studied long-term. Nat Rev Neurol 16:522

Losy J (2020) SARS-CoV-2 infection: symptoms of the nervous system and implications for therapy in neurological disorders. Neurol Ther 23:1–12

Meinhardt J, Radke J, Dittmayer C, Franz J et al (2021) Olfactory transmucosal SARS-CoV-2 invasion as a port of central nervous system entry in individuals with COVID-19. Nat Neurosci 24:168–175

Ramani A, Muller L, Ostermann PN, Gabriel E et al (2020) SARS-CoV-2 targets neurons of 3D human brain organoids. EMBO J 39:e106230

Rhea EM, Logsdon AF, Hansen KM, Williams LM et al (2020) The S1 protein of SARS-CoV-2 crosses the blood-brain barrier in mice. Nat Neurosci 24:368–378

# 3

# Saving or Replacing Nerve Cells: Which Strategy is More Successful?

As described in the previous chapter, it is not possible for most patients with neurodegenerative diseases to determine the exact pathomechanism that ultimately leads to neuronal cell death. Even if a gene defect is known, as in familial cases of neurodegeneration, a causal therapy is not yet possible. If a gene therapy were to be available in the future that could replace a diseased gene with the intact gene, the progression of Parkinson's or Alzheimer's might be prevented. However, a comprehensive gene therapy for the human brain is still in its infancy.

Both the pathomechanistic investigations described in the second chapter and the resulting attempts to treat neurodegenerative diseases were primarily carried out in cell cultures and experimental animals, mostly in mice or rats. However, humans and rodents have had their last common ancestor around 70 million years ago and have developed independently of each other since then. Due to this evolutionary divergence, a rodent model of a neurodegenerative disease can only be a vague approximation of human pathology, especially since the aging mechanisms are very different between species. The "inner clocks", which determine the aging process of cells, go completely different in rodents and humans.

The complexity of neurodegenerative diseases and the general difficulty of finding suitable animal models for testing new therapies have motivated scientists to intensify research in the field of human cellular models, in particular, regarding stem cells. Since it is rarely possible to directly remove neuronal stem cells from the human brain, human **induced** and **pluripotent**

L. P. Klimaschewski, *Parkinson's and Alzheimer's Today*, https://doi.org/10.1007/978-3-662-66369-1_3

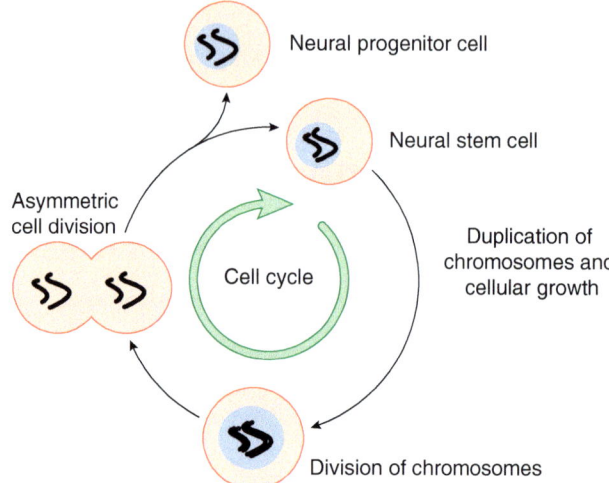

**Fig. 3.1** Neural stem cells are obtained from fetal brain tissue, cultured embryos (blastocysts), or by re-programming from adult, somatic body cells. They divide and generate progenitor cells which differentiate into a specific cell type only. Stem cells are multipotent and replicate indefinitely (progenitor cells divide only a limited number of times). After doubling their chromosomes, dissolving the cell nucleus, and forming two daughter cells, one cell re-enters the cell cycle and the other differentiate. By treatment with specific growth factors, postmitotic neurons and various types of glial cells are produced from progenitor cells that can be used for research purposes or transplantation experiments

stem cells (hiPSCs) have been produced in the laboratory to obtain neural cells for research and transplantation purposes (Fig. 3.1).

With this technology, stem cells can be obtained from skin biopsies or from the blood of patients with mutations in a Parkinson's or Alzheimer's gene. These cells are then **re-programmed**, i.e. differentiated into functional nerve cells, astrocytes, or microglia by introducing appropriate genes and treatment with specific growth factors.

For example, fibroblasts from Alzheimer's patients, which were converted into neurons, show an increased formation of proteins associated with dedifferentiation and cellular stress after detailed analysis of their genomes and comparison with induced stem cells from healthy controls. Through such findings, new hypotheses about the pathogenesis of Alzheimer's disease can be formulated and verified. In addition, known changes in gene expression, in lipid metabolism or endosomal transport can be investigated in such cells. The efficacy of drugs is already being tested on induced neurons or glia before they are taken by the patient. This saves costs and unpleasant side effects of a pharmacological therapy, which would perhaps not even help. By applying new gene-editing techniques (e.g. CRISPR-Cas9), the DNA of hiPSCs has already been changed to investigate the effects of rare Alzheimer mutations in cell culture, for example.

However, it must not be overlooked that the re-programmed cells are not in every respect comparable to the original neuronal and glial cells of the brain. For example, the induced neurons are usually immature, i.e. they resemble cells that are still in embryonic development. Also, changes that occur later in the course of a disease, such as the formation of tau fibrils in Alzheimer's disease, cannot be detected even though the cells are kept in culture for many weeks. Furthermore, the environment of the neurons in culture is very different from that in the brain, since the blood supply and the incorporation into a three-dimensional matrix with neighboring cells are lacking. Due to these considerable deficits of stem cell cultures, so-called brain organoids were developed in a further step. These brain-like three-dimensional structures arise from stem cells in cell culture under certain conditions. However, they also represent an early, non-vascularized stage of development and the typical age-related changes of neurodegenerative diseases are not present either.

However, the current state of research makes it possible to slow down neuronal cell death in animal experiments and to replace dead cells with stem cells added exogenously. These cells then differentiate in the brain either into neurons, which may even be integrated into existing neuronal networks, or into glial cells. Unfortunately, however, the clinical studies carried out so far in humans have not been successful in this respect. Therefore, the main focus nowadays is on trying to stimulate the normally low intrinsic (endogenous) neurogenesis pharmacologically. First experiments with growth factors and special drugs that promote cell division in resting stem cells in the brain were promising. They are all the more important as there are currently no neuroprotective or disease-modifying therapies available. The paragraphs 3.1.6 and 3.2.5 in this chapter will go into more detail on clinical stem cell studies for the treatment of Parkinson's and Alzheimer's disease.

In the following, I will first introduce the common symptomatic therapies of Parkinson's and Alzheimer's disease and then discuss the experimental approaches to treat or prevent neuronal degeneration.

# 3.1    Parkinson's Disease

In general, it must be distinguished in the therapy of Parkinson's disease whether the treatment promotes the survival of nerve cells or replaces the missing transmitter dopamine. The latter is the primary goal of the pharmacological symptomatic therapy currently in use. In addition to the motor functions, autonomic and cognitive or psychiatric symptoms may also be improved pharmacologically. Furthermore, surgical (stereotactic) interventions and physio- or speech-therapy are available.

## 3.1.1  Pharmacological Therapy

Interestingly, neuroprotective, survival promoting effects of the dopamine—receptor agonists, which were developed 50 years ago and intended to simulate the transmitter dopamine only, had been suggested for some time. For example, bromocriptine (an ergotamine derivative and agonist at the D2 receptor), pramipexole or ropinirole (agonists at the D2/D3 receptor) and R-apomorphine (agonist at the D1/D2 receptor) not only imitate the dopaminergic neurotransmission, but contribute to a reduction in neuronal cell death in the substantia nigra in animal models. The substances mentioned are given in particular to younger patients before a L-dopa therapy, since they cause less motor complications (e.g. excessive muscle activity). But they can also cause hallucinations, especially in older patients with cognitive deficits.

L-Dopa is currently the most effective Parkinson's medication. As a precursor to dopamine, it diffuses through the blood-brain barrier and is converted to dopamine in the brain (Fig. 2.5). Dopamine cannot, after all, overcome the barrier orally or intravenously. Treatment is therefore carried out together with a DOPA-decarboxylase-inhibitor (benserazid), which does not enter the central nervous system, but prevents the unwanted conversion of L-Dopa to dopamine outside the brain.

Unfortunately, the effectiveness of L-dopa decreases significantly over the years. The side effects of prolonged administration can also be considerable. If the dopaminergic effect is too strong, overshooting movements (hyperkinesia) can be observed. Hallucinations, restlessness or anxiety disorders may occur as well. Most often, the L-dopa dose is then reduced or the therapy is combined with inhibitors of enzymes that degrade dopamine, for example with COMT or MAO-B inhibitors (rasagiline). The

Catechol-O-Methyl-Transferase (COMT) and the Monoaminoxidase B (MAO-B) metabolize L-dopa and dopamine (Fig. 2.5). Their inactivation therefore leads, analogously to the therapy with L-dopa, to higher dopamine levels in the brain.

In addition to these drugs directly modulating the dopaminergic system, amantadine, a blocker of glutamatergic activation at the NMDA receptor, is also used in Parkinson's patients. Glutamate as the most important activating transmitter in the brain can, as discussed in the second chapter, lead to neuronal cell death via overexciting neurons. Therefore, a glutamate receptor blockade has a favorable effect on the survival of nerve cells under certain pathological conditions. However, the effects of glutamate in the brain may not be generally inhibited. Here, the problem of most drugs used in neuropsychiatric disorders becomes apparent. They just flood the whole brain during systemic treatment, although they should only be effective in very specific brain areas. This fact usually prevents an effective dosage of the drug, which would be necessary to generate the necessary effect, as otherwise too many side-effects occur.

The lack of motor activity in Parkinson's disease may also be positively influenced by an inhibition of glutamatergic transmission in particularly active neurons of the nucleus pallidus internus or nucleus subthalamicus. These nuclei are themselves movement-inhibiting (see Fig. 2.8), and in terms of a double-inhibition, such treatment would lead to an improvement in symptoms. In addition, anti-cholinergic drugs (e.g. trihexyphenidyl) are used for their relaxing effect on muscles in cases of severe tremor.

## 3.1.2 Surgical and physical therapy

In addition to pharmacological treatment, surgical procedures are available to Parkinson's patients. It can be said that the deep brain stimulation has revolutionized the treatment of movement disorders. By 2020, 140,000 mostly younger patients had already been treated with such a brain pacemaker. For this purpose, electrodes are stereotactically introduced into different nuclei of the diencephalon. A metal frame is firmly attached to the patient's head and allows precise placement of the electrodes in defined regions. With the nucleus subthalamicus, the globus pallidus internus and the nucleus ventralis intermedius of the thalamus, three different target areas are available for Parkinson patients, the latter nucleus especially for the treatment of tremor. After insertion, the electrodes are connected to an electrical stimulator that is implanted under the skin in the area of the clavicle.

In most patients treated in this way, positive effects are already visible after a short time, which are more pronounced than those achievable with pharmacotherapy. The tremor subsides and mobility improves. Often, more than half of the medication taken can be saved. In addition, good physical mobility is achieved throughout the day in some of the patients. However, the electrodes erode slowly and their position may change. In addition, the batteries for the control unit have to be replaced surgically. Therefore, new ultrasound-based methods are currently being developed to switch off defined brain nuclei non-invasively under magnetic resonance imaging (MRI) control.

Physiotherapy to prevent secondary joint stiffness is useful for practically all patients. Various forms of physical activity, including ergometer training, not only improve motor symptoms, but also reduce depressive mood and anxiety, as well as sleep and memory problems. Regular gait training, for example Nordic walking, stabilizes direct cortico-spinal connections, so that the defective basal ganglia are skipped during the activity. In addition, practice of arm movements during hiking is beneficial. For patients with speech disorders (dysarthria), voice training is recommended, which can also improve speech loudness. For alternative therapies, such as the cowhage used in ayurvedic medicine (Mucuna pruriens), acupuncture, manual therapies (chiropractic, massage, etc.) or biofeedback, efficacy data from controlled clinical studies are still lacking.

### In a Nutshell

- The current therapy for Parkinson's disease is symptomatic, as there are no verified procedures available today that can ensure the survival of dopaminergic neurons or replace degenerated cells.
- L-Dopa is the most effective Parkinson's medication currently available. It is given in combination with a DOPA decarboxylase inhibitor to prevent the conversion of L-Dopa to dopamine outside the CNS.
- With stereotactic surgery, all three cardinal symptoms of the disease (rigor, tremor, akinesia) can be improved in severe cases.
- Deep brain stimulation alters the activity of neurons in the subthalamic nucleus, the internal globus pallidus, or the ventral intermediate nucleus of the thalamus via electrodes surgically introduced by neurosurgeons.

### 3.1.3  Therapy with Neurotrophic Factors

In the 1980s, scientists proposed treating neurodegenerative diseases with neurotrophic molecules in order to prevent or at least delay the neuronal cell death. In addition to the proteins of the family of neurotrophic factors, smaller, 20–40 amino acid-long neuropeptides, such as PACAP (pituitary adenylate cyclase activating polypeptide) or galanin, promote the survival of neurons. Both peptides activate G protein-coupled receptors in the neuronal plasma membrane (Fig. 3.2). The neurotrophins, on the other hand, bind to receptor tyrosine kinases (RTKs). Both classes of receptors positively

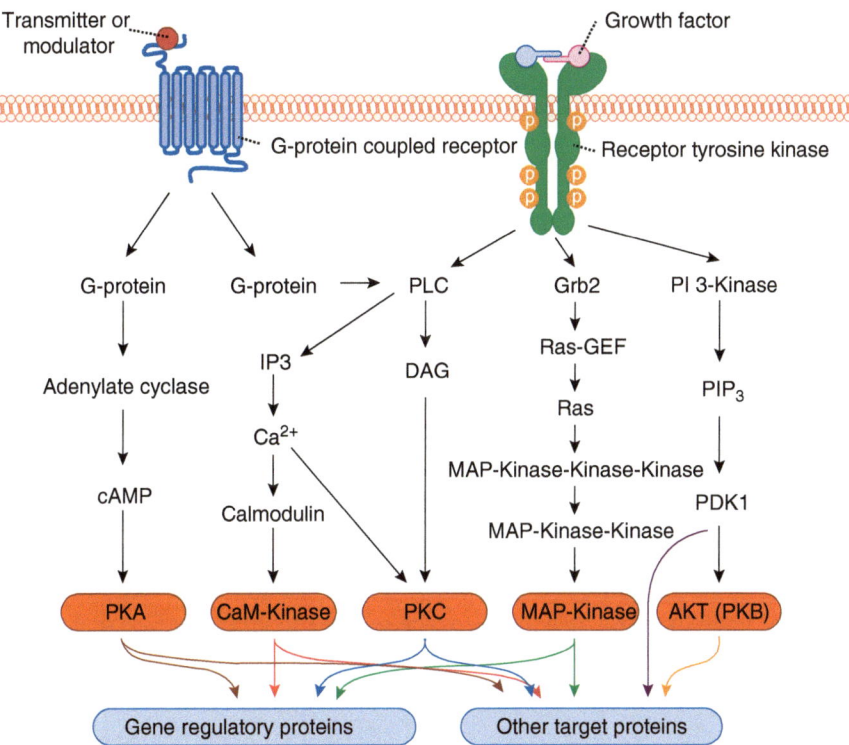

**Fig. 3.2**  Intracellular signaling pathways that originate from transmitter and growth factor receptors. Receptor tyrosine kinases (green) are usually activated by dimerization, i.e. one receptor molecule binds a growth factor, the receptor ligand, and interacts with a second ligand-receptor complex in the plasma membrane. The receptor kinase domains protruding into the cytoplasm then phosphorylate each other and activate various pathways: the PLC/DAG/PKC, the Grb2/Ras/Raf/MAP kinase (ERK) and the PI3 kinase/AKT signal transduction cascades. G protein-coupled receptors representing, for example, the dopamine receptors, consist of a membrane protein (blue) to which a G protein docks. This binds GTP and activates adenylyl cyclase, which forms the secondary messenger cAMP and thus activates protein kinase A (PKA). In addition, various $Ca^{2+}$-dependent signaling pathways are switched on via phospholipase C (PLC)

influence the growth of axons and dendrites, too. Long-term neuronal changes induced by trophic factors are usually mediated via the regulation of gene expression in the neuronal nucleus, i.e. via altered production (expression) of mRNAs.

The RTKs activated by neurotrophic factors stimulate in particular the PI3-kinase/AKT- and Ras/Raf/MAP-kinase-dependent signaling pathways. PI3-kinse dependent mechanisms are essential for neuronal survival and axonal growth. Members of the neurotrophin family (NGF, BDNF, NT-3) and of neurotrophic cytokines, such as the ciliary neurotrophic factor (CNTF), have shown impressive effects in animal models of neurodegenerative diseases. Similarly, insulin-dependent growth factors (IGF-1 and IGF-2), the transforming growth factors (TGFs) and various members of the fibroblast growth factor family (FGFs) have been found to be protective as well. These molecules are mainly released by glial cells. However, nerve cells themselves also produce neurotrophic factors, including the brain derived neurotrophic factor (BDNF). BDNF helps neurons to keep themselves alive (via autocrine effects), but also affects neighboring cells (paracrine effect) and promotes dendritic growth (morphological plasticity).

In neurological diseases associated with neuronal degeneration, neurotrophic factors are hardly ever found in the affected brain regions. Therefore, pharmacological approaches are currently being pursued to activate RTKs via alternative mechanisms and enhance receptor levels in neurons by promoting their recycling or inhibiting their degradation in the lysosome. Since the relevant signal transduction pathways are also controlled by receptors localized in endosomes, the delay of transport of activated receptors from Rab5-positive early endosomes to Rab7-positive late endosomes prolongs their activity intracellularly. Moreover, attempts are being made to increase the transfer of intracellular RTKs to Rab11-containing recycling endosomes. This way, growth factor receptors can be re-used in a cell and made available again for pharmacological treatment with receptor agonists (Fig. 3.3).

The ability to artificially produce neurotrophic molecules in large quantities was a driving force for the biotechnology industry that emerged in the 1980s to treat patients with Parkinson's or Alzheimer's disease. Unfortunately, almost all of these treatment attempts failed. The reasons for that failure are not entirely clear. First of all, it was not yet known at that

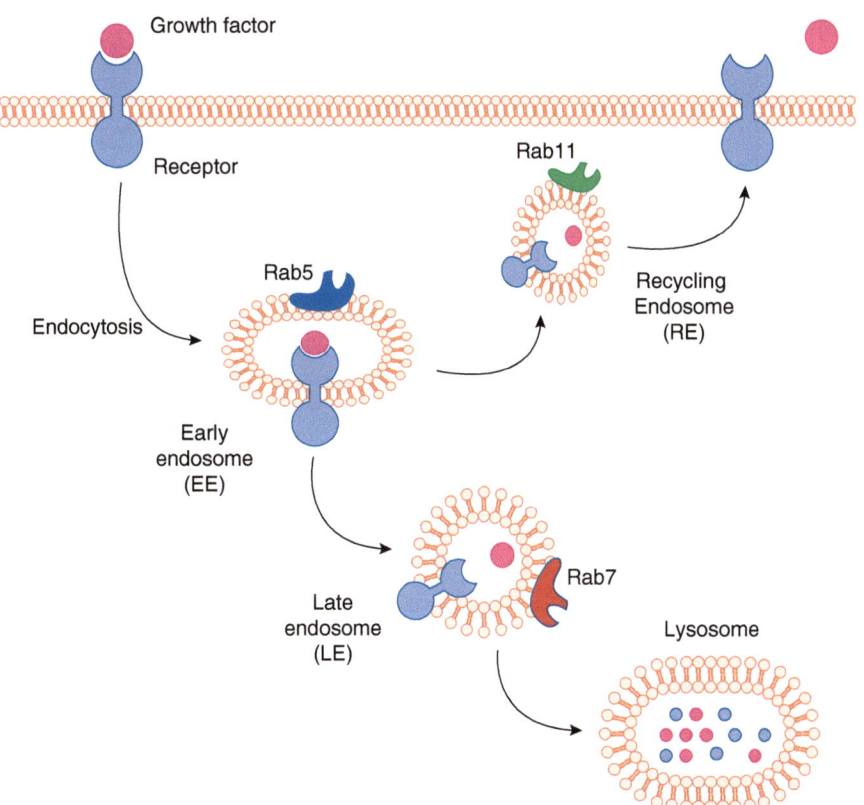

**Fig. 3.3** Early and late endosomes can be distinguished by membrane-bound markers (Rab GTPases, Rab5 in early, Rab7 in late endosomes). If the endosomes return to the cell surface and fuse again with the plasma membrane, they are referred to as recycling endosomes. Early endosomes are those that form immediately after endocytosis. Intraluminal vesicles can be formed by invagination of membranes into late endosomes. Endosomes containing smaller endosomes are called multivesicular bodies (MVBs) (see also Fig. 2.3)

time in what dosage and in what form the trophic factors had to be applied in order to reach the brain tissue in sufficient quantities. Unfortunately, this was usually not the case.

In addition, there were no studies on the question of which effects a certain factor would have in the different brain areas exactly. Probably they would have had to be administered in higher doses or stereotactically (that is, by means of a probe) directly into specific brain regions. When given into the blood or even into the cerebrospinal fluid (CSF), proteins are quickly diluted and usually degraded within hours to days. Even with injection of high concentrations, the factors would therefore be inactivated by protein hydrolysing enzymes (peptidases or proteases) that circulate in the serum and cerebrospinal fluid. It is still being discussed today whether neurons in the brain of the elderly express sufficient amounts of intact receptors (e.g. RTKs). The growth factors have therefore not yet found their way into the regular therapy of Alzheimer's and Parkinson's disease outside of clinical studies.

In regard to Parkinson's disease, the most frequently investigated growth factor is glia-derived-neurotrophic factor (GDNF). GDNF promotes the survival and maturation of dopaminergic neurons, but was not successful in the first clinical studies using direct application of the protein. In more recent approaches, GDNF was introduced into the brain by gene therapy. Other therapeutic attempts involve newly identified trophic molecules such as cerebral dopamine neurotrophic factor (CDNF). CDNF is experimentally effective, although it does not bind to surface receptors, but apparently can positively influence ER stress in neurons by binding to intracellular receptors in the endoplasmic reticulum (ER). In this context, the inhibition of the integrated stress response (ISR) is of therapeutic interest as well, because the reduction of neuronal protein synthesis in the context of cellular stress promotes neuronal degeneration.

### 3.1.4   Therapy with Antisense Oligonucleotides

Pathological proteins, for example mutated α-synuclein, can be recognized by antibodies and eliminated by the immune system. An alternative mechanism to interfere with unwanted proteins would be to reduce or completely suppress their endogenous expression. In children with spinal muscular atrophy, for example, a therapy with antisense oligonucleotides (ASOs) has been approved for the first time (SPINRAZA, Biogen). Further clinical studies with promising ASOs have been started in patients with Huntington's disease, amyotrophic lateral sclerosis and also in Parkinson's disease.

ASOs are synthetic, 16–22 base long DNA molecules in the form of a single strand (the DNA in the cell nucleus is double-stranded). They selectively bind to the targeted mRNA by complementary base pairing and thus prevent the formation of the corresponding protein on the ribosomes (Fig. 3.4). Moreover, ASOs activate RNA-cleaving enzymes (RNases) or block the initial mRNA triplets (the start codon) at which protein synthesis begins. In addition, they can interfere with the processing of precursor mRNAs

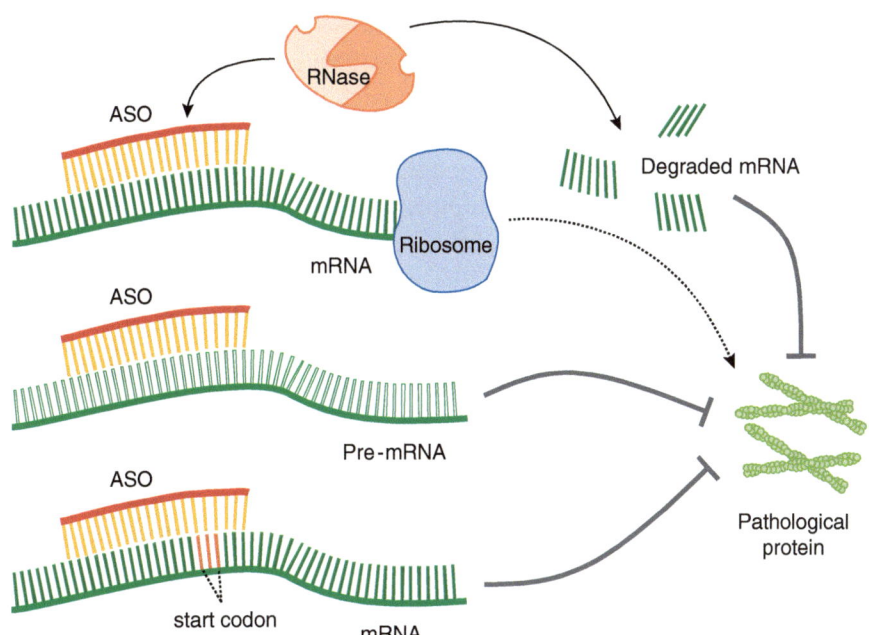

**Fig. 3.4** Reduction of pathological proteins by treatment with antisense oligonucleotides (ASOs). These bind to complementary mRNA or precursors of mRNA (pre-mRNA) which results in the degradation of mRNA by RNases, so that no templates are available for the ribosome. Ribosomes produce the target protein on the basis of the template sequence (dotted arrow). The binding of ASOs to pre-mRNA prevents its processing into mature mRNA and thus also protein synthesis. The superposition of the ASO with the start codon, at which the ribosome would initiate the production of the protein, does not allow protein production at the ribosome either, so that the amount of the protein produced decreases

(pre-mRNAs) in the cellular nucleus and modify the RNA splicing mentioned above.

ASOs have been in experimental use for a long time, especially in cell culture. However, their rapid degradation by cellular nucleases has so far prevented their use in humans. However, due to various chemical modifications, their half-life can be significantly extended and therapeutic successes were achieved. These modifications include the phosphorothio- and the 20-O-methoxyethyl (20-MOE)-sugar modifications of the DNA, which make ASOs water-soluble and more resistant to cleaving enzymes (exonucleases).

In principle, however, the *small interfering RNAs* (siRNAs) could also be used to block protein synthesis. These are short, only 20–25 base long RNAs that, analogously to the ASOs, enter into a complementary binding with the mRNA and thus prevent the corresponding protein from being produced. In contrast to siRNAs, however, ASOs are more easily taken up by neurons and show more dose-dependent and reversible effects. Furthermore, ASOs do not use the same RNA-processing mechanisms of cellular metabolism as siRNAs. An ASO therapy is therefore less susceptible to negative side effects. However, one disadvantage of ASOs is that they have to be given more often, as they have a shorter half-life than the siRNAs. Implantable pumps could possibly help, and viral vectors would also be a possibility to allow ASOs to be produced by nerve cells over a longer period of time.

ASOs were developed, for example, against leucine-rich repeat kinase 2 (LRRK2). This is an enzyme that lies at the root of most hereditary forms of Parkinson's disease. Mostly *gain-of-function* mutations of LRRK2 have been described, i.e. its activity is not reduced by the mutation but increased. Perhaps LRRK2 normally inhibits autophagy and thus contributes to an intracellular accumulation of proteins. Since its kinase domain is accessible to drugs, corresponding kinase inhibitors are in preparation (e.g. GNE-7915). Unfortunately, GNE-7915 causes unwanted side effects in various organs. LRKK2-ASOs would therefore be advantageous. To test the ASO mediated down-regulation of LRKK1, fibrillar α-synuclein was administered into the striatum of mice which caused degeneration of the substantia nigra and thus a Parkinson's disease-like syndrome. After injection of the LRKK2-ASOs directly into the brain of these mice, the kinase levels and the α-synuclein aggregates were significantly reduced. Many neurons in the midbrain could be saved in this way and the failures in motor skills of the mice reduced. These results are very interesting and give patients hope that interference with LRRK2 should also be effective in sporadic Parkinson's disease. As always with such preclinical studies, however, an independent repetition of the animal experiment and a tolerability test in humans must be carried out before Parkinson's patients can be treated with LRRK2-ASOs in clinical studies. The company Biogen has started such study in August 2019 (https://clinicaltrials.gov/ct2/show/NCT03976349).

Particularly promising appears to be the shutdown of an RNA-binding protein (PTB) which is down-regulated after the completion of neurogenesis during development. PTB prevents neuronal differentiation of precursor cells and the production of typical neuronal proteins such as tubulin III (Tuj1) or the dendritic MAP2. The injection of ASOs against PTB into the substantia nigra of mice leads to the transformation of astrocytes into dopaminergic neurons, whose axons grow into the striatum and actually release dopamine there. Intriguingly, the motor deficits of the animals were also corrected by this treatment, so that not only a putative new Parkinson's therapy, but also a promising strategy for the regeneration of nerve cells in the aging brain was described in this study for the first time.

### 3.1.5 Alpha-synuclein Aggregation Inhibitors and Specific Immunotherapy

As discussed in the previous chapter, α-synuclein can be passed on from one cell to another in the brain. Hence, it should be possible to bind extra-cellular α-synuclein with antibodies, so that it is cleared by microglia or by cells of our general immune system. Based on this idea, vaccines have been developed for both α-synuclein and Alzheimer's amyloid (see Sect. 3.2.4) in the form of an active immunization. On the other hand, humanized antibodies can be injected (passive immunization). The possible success of these approaches needs to be checked in clinical studies. Furthermore, molecules are tested that reduce the aggregation of α-synuclein (e.g. NPT200-11). In mice that produce too much α-synuclein and thus develop Parkinson-like symptoms, therapeutic effects of NPT200-11 have already been demonstrated.

Various low-molecular substances (NPT200-11, CLR01, SC-D, ZPD-2) can interact with α-synuclein and reduce its neurotoxic effects in cell culture as well as in animal experiments. This is probably due to the prevention of the formation of oligomers or the dissolution of already existing fibrils. SC-D and ZPD-2 also prevent the aggregation of α-synuclein mutants (H50Q, A30P) that cause an early form of Parkinson's disease. However, none of these molecules, also referred to as α-synuclein stabilizers, have yet made the jump into the clinic.

### 3.1.6 Stem Cell Therapy

In light of the fact that Parkinson's disease is characterized by a well-defined population of dopaminergic neurons in the midbrain, great hopes have been placed in the cell replacement therapy outlined above. As early as the 1990s, researchers began transplanting precursor cells from fetal human midbrain into the basal ganglia, specifically into the striatum. In some Parkinson's patients, improvements in symptomatology were indeed detectable afterwards, but most patients did not benefit from this approach in double-blind controlled studies (i.e. neither the doctor nor the patient knew whether stem cells were present in the syringes used for injection). In addition, an intensive discussion began about the ethical and practical problems posed by the use of aborted fetuses to obtain the stem cells. Consequently, alternative sources of transplantable cells were sought more intensively and resulted in the development of the pluripotent stem cell technology mentioned above for production of hiPSCs that can be converted into human dopaminergic neurons and are already used in clinical studies.

Stem cells can be either derived from embryos or obtained from adult cells by re-programming as described above (Fig. 3.1). By using inhibitors of the TGF/SMAD signaling pathway an almost unlimited number of typical dopaminergic neurons can be produced. If these are transplanted into the substantia nigra of monkeys, the new neurons are able to extend axonal projections into the striatum, where they release dopamine and positively influence motor behavior.

However, certain risks of this approach for patients need to be considered. It can not be excluded, for example, that transplanted cells divide further in the brain and thus may form tumors or teratomas, i.e. mixed tissue derived from stem cells. Their possible rejection by the human immune system is also not sufficiently investigated, because the blood-brain barrier opens up during surgery and immune-competent cells could gain access to the transplanted cells. This problem could possibly be solved by using HLA-compatible hiPSC donors (analogous to bone marrow transplantations). However, a number of symptoms that are associated with the reduction of other transmitters in Parkinson's patients (e.g. dementia, weight loss, smell and speech disorders) can not be treated by introducing dopaminergic neurons into the basal ganglia.

Another important aspect of therapy with stem cell-derived neurons concerns the exact mechanism of their action. Are the positive effects found in experimental studies due to the integration of these cells into existing neuronal networks? Or is the release of dopamine and growth factors from the transplanted cells sufficient to produce the desired effects? To answer these questions, several experimental studies have been conducted, which should be summarized briefly here.

In animal experiments with fetal midbrain cells, synaptic contacts were observed both from transplanted cells to striatal neurons of the recipient and, vice versa, from recipient neurons to the transplant. Such synapses were also detected after transplantation between recepient neurons and dopaminergic neurons derived from hiPSCs. The neuronal contact sites first appeared 1–2 months after transplantation and were detectable for at least 6 months. We must therefore assume that stem cells can differentiate into neurons in the brain and are capable of integration into neuronal networks.

Finally, the neuronal activity of transplanted cells can be turned on and off by optogenetic stimulation with light pulses. Since the motor symptoms are influenced in the expected way by this technical trick, it can be assumed that hiPSCs are built into existing motor networks and take over the function of the degenerated dopaminergic brainstem neurons in these challenging animal experiments.

## 3.1.7  Other Causal Therapeutic Approaches

In the search for disease-modifying therapies for Parkinson's disease, drugs used to treat lysosomal storage diseases have been in the focus in recent years. Lysosomal disorders often go hand in hand with neurodegeneration and drugs that improve the endosomal-lysosomal transport of proteins could provide new therapeutic avenues for Parkinsons's disease. Other therapeutic attempts with calcium channel blockers, anti-oxidants such as coenzyme Q10 or zonisamide, which increases GSH levels in astrocytes, or an improved supply of neurons with ATP (e.g. via creatine phosphate) were not effective in clinical studies so far. There is neither sufficient evidence for therapeutic effects by changing our eating habits. Recommendations have been made in the direction of a ketogenic diet with an increased intake of medium-chain fatty acids or anti-oxidative and anti-inflammatory substances, such as coconut oil, curcumin, vitamins or flavonoids. So far, none of these approaches has been successful in controlled clinical studies.

**In a Nutshell**

- Neurotrophic proteins and peptides have been studied for many years with regard to their possible promotion of the survival of aging and diseased neurons.
- These molecules activate receptor tyrosine kinases or G protein-coupled receptors, which delay or prevent neuronal cell death through defined signal transduction pathways in experimental models of Parkinson's disease.
- Pathological α-synuclein is recognized by antibodies and eliminated by the immune system. Therefore, vaccinations against this protein are being tested.
- In addition, disease-associated proteins may be targeted by antisense oligonucleotides (ASOs) or small interfering RNA oligonucleotides (siRNAs).
- Possible candidates for such a therapy would be leucine-rich repeat kinase 2 (LRRK2) or an RNA-binding protein (PTB) in Parkinson's patients. Reducing PTB in astrocytes of the midbrain creates nerve cells that project into the striatum and release dopamine there.
- Pluripotent and inducible stem cells can be obtained from embryos or from adult cells by re-programming strategies.

- After conversion into neurons and transplantation into the striatum or the substantia nigra, the cells release dopamine and growth factors. They partly also integrate into existing neuronal networks and have a positive effect on motor functions.

## 3.2   Dementia and Alzheimer's Disease

To date, there is no convincing therapy for dementia available, although some medications that enhance the effect of transmitters such as acetylcholine may lead to a temporary improvement of symptoms. Unfortunately, the immunization attempts against Aβ-Peptide have not shown the desired effects in clinical studies, although they had been successful in animal models. Neither is Alzheimer's disease slowed down by a therapy directed against amyloid, nor does it stop completely.

We must therefore assume that the Aβ pathology is not the decisive or even the sole cause of the death of neurons and the resulting mental decline in Alzheimer's. Rather, it seems to be a combination of several factors: Tau fibrils, gliosis (astrocytosis), toxicity by free radicals and circulatory problems are likely to be jointly responsible for the development of the disease. Thus a successful treatment will have to be directed against several factors. Probably, in the context of such a combination therapy, the cognitive enhancers will also play a role. These are substances that influence aminergic transmitter systems, for example, by blocking postsynaptic serotonin receptors (5-HT-6 or 5-HT-3).

### 3.2.1  Cholinergica

As discussed in the previous chapter, Alzheimer's disease is primarily characterized by a deficiency of acetylcholine, an important neurotransmitter that is reduced in the neocortex at an early stage of the disease (Fig. 2.15). The cortex is supplied by cholinergic neurons that are located in the basal forebrain. In Alzheimer's patients, the enzymes that produce acetylcholine (choline acetyltransferase) or degrade it (cholinesterase) are reduced. The remaining esterase activity can be pharmacologically inhibited. The esterase inhibitors are therefore theoretically useful drugs that delay the breakdown of acetylcholine in the brain and thus increase the concentration of this neurotransmitter at the synaptic contacts. Substances that irreversibly bind to the active center of acetylcholinesterase are toxic in very small amounts and are used as chemical weapons (e.g. tabun, sarin or novichok).

The reversible cholinesterase inhibitors donepezil or galantamine are prescribed for mild to moderate Alzheimer's dementia. In addition, galantamine (isolated from snowdrops) can also bind to acetylcholine receptors and thus increase their sensitivity. Improvements in cognitive functions were be demonstrated for both substances. However, these are only temporary and too weak for a noticeable improvement in the quality of life for most patients. Although a statistically significant effect was demonstrated in several studies and the esterase inhibitors were therefore "evidence-based" approved, their recommendation is questionable. This is a general problem in the prescription of some medications that show statistically significant effects in controlled studies with thousands of patients, but in relation to side effects and costs are not sufficiently large to justify their widespread use.

Other therapeutic approaches have attempted to enhance acetylcholine levels in the brain by increasing the synthesis of acetyl-CoA and choline. Furthermore, choline reuptake could be stimulated. In addition, the release of acetylcholine at the synapse or the sensitivity of postsynaptic binding sites, i.e., nicotinic or muscarinic receptors, may be promoted experimentally. So far, however, these strategies have not found their way into the treatment of Alzheimer's disease patients.

In addition to improving the cholinergic system, medication can also influence other transmitters in order to, for example, increase the effects of dopamine or serotonin. This is intended to treat the neuropsychological symptoms that occur alongside the memory loss and are likely caused by a deficiency of other biogenic amines. Finally, neuronal over-excitation caused by glutamate, which is released in large quantities by degenerating neurons, can be treated with NMDA receptor antagonists (memantine).

The following sections will introduce therapeutic strategies not yet available for all Alzheimer's patients, but are experimental or tested in clinical studies.

## 3.2.2  Therapy with Secretase Inhibitors

An important pharmacological target is the processing of the amyloid precursor protein APP by membrane-bound enzymes described above (see Fig. 2.17). The γ-secretase produces Aβ-peptides and can be blocked by γ-secretase inhibitors (GSIs). GSIs lead to a decrease in pathogenic peptides, which essentially form the amyloid plaques. However, the γ-secretase is also essential for cell differentiation via the Notch signaling pathway. A complete systemic blockade of this enzyme is therefore not possible.

Significant side effects are also observed with inhibitors against β-secretase (BACE1), which in this case are explained by the unspecificity of the chemicals, since other proteases important for cell metabolism are also inhibited. In addition, BACE1 has important functions in the brain, e.g. in the myelination of axons and in the maintenance of neuronal networks. Moreover, the enzyme is involved in the formation of dendritic spines, which are decisive for memory formation. BACE1 inhibitors have therefore been abandoned.

The development of γ-secretase modulators (GSMs) that reduce the levels of pathological Aβ42/43 peptides, but have fewer side effects, is considered particularly promising. GSMs may be used early on in patients with autosomal-dominant familial presenilin mutations. However, the first clinical studies have been disappointing so far, but research is ongoing.

### 3.2.3  Therapy with Neurotrophic Factors

In the previous section, the general importance of growth factors for the treatment of neurodegenerative diseases was discussed. It was known early on that the nerve growth factor (NGF) not only keeps peripheral neurons alive, but promotes survival of central neurons located in the brain as well, e.g. in the nucleus basalis (the cholinergic nucleus located in the ventral forebrain). These neurons, which are affected early on in Alzheimer's disease, can be saved after their axons have been cut by NGF treatment.

All neurotrophins, but also the IGFs, FGFs or some of the bone morphogenetic proteins (BMPs), act in the mature brain mainly through an increased activation of PI3K/AKT-dependent signaling pathways in neurons bearing the respective receptors (Fig. 3.2). This can, for example, counter an increased p38 activity, which leads to neuronal apoptosis. Since AKT also inhibits the activity of the enzyme GSK3β important for Tau phosphorylation, neurotrophic factors also reduce the hyperphosphorylation of Tau. The Aβ oligomers discussed in the second chapter lead to an increased activation of GSK3β, which in turn results in an activation of glutamatergic receptors, the NMDA ion channels. This allows more cations to flow into axons, thereby disturbing synaptic transmission. Neurotrophic factors counteract this effect. Furthermore, it has been observed in cell culture models of Alzheimer's disease that the activities of protein kinase C (PKC), Jun kinase (JNK) and Wnt/β-catenin dependent signaling pathways are generally low, which contributes to synaptic dysregulation.

Of the neurotrophic factors, basic fibroblast growth factor in particular shows promising effects in a mouse model of Alzheimer's disease. Subcutaneous administration of FGF2 alleviated the deficits in spatial memory in the animals and reduced the aggregation of Aβ and Tau. At the same time, activity of BACE1 was inhibited, so that fewer Aβ peptides were formed by this secretase. Since FGF2 is produced in small amounts only in the brains of Alzheimer's patients, clinical studies with FGF2 are warranted.

In order to avoid the above-mentioned problems of treatment with growth factors, i.e. their rapid degradation and the low penetration through the blood-brain barrier, short peptides 20–40 amino acids in length have been developed which can more easily overcome the blood-brain barrier than large proteins. These peptides activate—just like the naturally occurring proteins—the corresponding receptors (RTKs) and thus stimulate the neurotrophic intracellular signaling pathways. In addition, non-peptidergic and therefore often blood-brain barrier permeable receptor agonists would be of interest. So far, all of these approaches are still in the experimental stage. It is also not clear whether the RTKs in the membrane of aging neurons can forward signals as effectively as in the young brain. Therefore, some laboratories (including my own) intend to increase the occurrence of functional RTKs in aging neurons by inhibiting their degradation or promoting receptor recycling.

In this context, special attention should also be paid to possible effects of growth factors on glial cells. Astrocytes often show atrophic changes in neurodegenerative diseases. Like nerve cells, they express receptors for neurotrophins and FGFs. These factors, but interestingly also physical activity or an enriched environment, stimulate astrocytes in the brain. They show a higher morphological complexity with longer cell processes. In addition, the drainage of the cerebrospinal fluid via the glymphatic system is promoted, since activated astroglia transports more water channels (aquaporin 4) into its processes contacting blood vessels.

## 3.2.4 Immunotherapy

In recent years, a number of antibodies against Aβ-peptides have been developed, which are mainly responsible for the formation of Alzheimer's plaques. Some of the antibodies directed against Aβ were first discovered in healthy people who never developed Alzheimer's dementia, even in old age. They were thus intrinsically immunized without ever having been vaccinated.

These antibodies were analyzed in more detail and produced for clinical trials in humans. In addition, non-immunologically active substances against Aβ monomers, oligomers or amyloid aggregates were developed.

The clearing of Aβ, the pharmacologically stimulated transport of Aβ peptides in the brain, is in the focus of many researchers. Some of these approaches show partial efficacy in animal studies preceding each clinical study, i.e. they lead to a reduction of Aβ oligomers and to the dissolution of amyloid plaques. Unfortunately, the clinical studies with these new candidates were unsuccessful. Apparently, the neuropathological changes, including the loss of synaptic contacts between neurons and the strong proliferation of glial cells (astrocytosis), are irreversible at the time of treatment.

Despite these depressing results, it was found on repeated analysis of the data that, after administration of high doses of newly developed antibodies directed against Aβ oligomers and smaller protein aggregates, improvements in memory function could be detected in some patients at early stage of Alzheimer's dementia. Perhaps antibodies such as Aducanumab (now approved in the USA) or Lecanemab (BAN2401) slow down the cognitive decline and reduce the amount of senile plaques at higher concentrations after all. It appears that the precursors of amyloid aggregates, the soluble Aβ oligomers, may, in fact, play a role in neuronal degeneration. However, it must be taken into account that, after intravenous administration of antibodies, only small amounts of immunoglobulins reach the brain (about 0.1% with an intact blood-brain barrier). Nevertheless, under pathological conditions the antibody uptake could increase, since the barrier function of the endothelium is reduced.

Clinical studies are also being conducted with antibodies directed against intracellular tau aggregates that can be taken up by neurons via Fc receptors (a form of passive immunization). Active immunizations against tau are being carried out as well in order to elicit an endogenous response against pathological tau. Furthermore, other approaches may be helpful that early on block the transmission of pathological tau from the brainstem into the forebrain, so that cortical degeneration will not occur. In addition, inhibitors of tau aggregation are being tested in humans. All of these strategies could be more effective than the Aβ-immunotherapy discussed above because the formation of tau fibrils correlates well with the clinical symptomatology and could also be used for other tau-associated neurodegenerative diseases, such as frontotemporal dementia. Finally, small molecules that may be used as inhibitors against tau modifications (acetylation, phosphorylation) or as microtubule stabilizers have been developed.

## 3.2.5  Stem Cell Therapy

Stem cell therapies are also discussed in the context of Alzheimer's disease, as in the case of Parkinson's disease. Today we know that in the hippocampus, which is located in the temporal lobe of the brain and undergoing atrophy in most Alzheimer's patients, new nerve cells may be formed. The hippocampal neurogenesis is present during development up to childhood and adolescence, but then gradually decreases. Although extremely rare in 90-year-old humans, a few newly formed nerve cells can usually be detected in the dentate gyrus of the hippocampus. Therefore, the idea is to stimulate neurogenesis in old age by the administration of growth factors that enhance proliferation and differentiation of endogenous stem cells. Alternatively, exogenous stem cell transplants could be used to replace degenerated neurons in highly affected areas such as the temporal lobe.

In this context, the already mentioned basic fibroblast growth factor (FGF2) is of major interest, since treatment with FGF2 by means of viral vectors into the brain of mice that produce too much APP as a model of Alzheimer's disease improved the spatial memory of the animals. In fact, an increased number of dividing neuronal stem cells and enhanced long-term potentiation (LTP) were observed. In addition, FGF2 can stimulate microglia to phagocytose Aβ peptides, which may also contribute to the reduction of symptoms after FGF2 administration.

## 3.2.6  Other Causal Therapeutic Approaches

As discussed in the previous chapter, it is assumed that Alzheimer's disease and other neurodegenerative diseases are initially caused by disturbances in the cellular energy balance or in the protein homeostasis. Apparently, the transport and degradation of proteins are disturbed early on. Currently, several strategies have been suggested that promote the intracellular transport of proteins. For example, molecules that increase the production of autophagosomes and lysosomes (e.g. TFEB), improve vesicle recycling (Rab11, Chmp2B) or stimulate lysosomal degradation (GBA, CatB) are considered as therapeutic targets.

Furthermore, certain protein complexes, for example the retromer, are of interest in this context, as they reduce the burden of the intracellular protein degradation machinery. The retromer promotes endosomal transport by returning membrane-bound molecules back to the cell membrane or to the Golgi fields. A receptor involved in this transport (sorting-related receptor with A-type repeats, SORLA) binds Aβ monomers or APP and thus helps to transport APP back to the plasma membrane and Aβ to the lysosomes. It is therefore not surprising that genetic variants of SORLA are associated with an increased risk of Alzheimer's disease.

Interestingly, a retromer-associated protein, VPS35 (vacuolar protein sorting-associated protein 35), leads to an autosomal-dominant inherited Parkinson's disease when mutated. These studies underscore the importance of endosomal protein transport in preventing neuronal damage. Selected molecules (R55, R33), which stabilize the retromer, reduce the formation of β-amyloid and phosphorylated tau in nerve cells. Unfortunately, R55 and R33 are quickly degraded in the blood and are therefore unsuitable for human therapy.

Furthermore, the novel inhibitors of the integrated stress response (ISR) would have to be mentioned here. They reduce the transient shut-down of neuronal protein synthesis as part of the cellular response to exogenous or intrinsic stress (e.g. in the case of glucose deficiency, hypoxia or ER stress). Recently, in a mouse model of Alzheimer's disease, it was shown that an ISR inhibitor, ISRIB, stimulates protein production in the hippocampus and improves synaptic plasticity and memory functions.

Another strategy involves the enhanced breakdown of proteins in the proteasome. After all, reduced proteasome activity can regularly be observed in neurodegenerative diseases. Various substances (betulinic acid from birch bark, pyrazoles or PD169316, a p38 MAP kinase inhibitor) increase the proteolytic activity of the proteasome and partially lower the intracellular concentration of α-synuclein in cells that produce too much of it. In the future, special antibodies (so-called *protein degraders*) could be used instead of the less permeable ASOs. With this novel approach, a protein produced can be marked for degradation by coupling it with ubiquitin.

The above-mentioned antisense technology may also be suitable for Alzheimer's disease in order to shut down or at least reduce the synthesis of specific proteins. ASOs directed against the mRNA of disease-associated proteins can temporarily stop or reduce the production of these proteins. There are already promising antisense approaches in animal models of Alzheimer's disease that can halt nerve cell death and improve memory functions.

Furthermore, molecules of interest are those that improve the function of lipid transporters. Since ApoE4, which is associated with Alzheimer's disease, may not be correctly folded, it can no longer sufficiently take over the transport of lipids and Aβ peptides. Therefore, newly developed substances

(GIND25 or PH002) that stabilize ApoE are particularly promising in ApoE4-positive Alzheimer's patients.

Finally, novel methodological aspects need to be discussed, as there have been significant advances in the development of biomaterials and carrier substances for active ingredients in recent years. These were produced to support the transport of growth factors, pharmaceuticals, ASOs, antibodies, but also of stem cells into the brain. Cell carriers, biological scaffolds, and in particular the currently intensively researched nano- and microparticles can bind or enclose therapeutic molecules. This prevents their degradation and promotes transport across the blood-brain barrier. In addition, attempts are currently being made in various laboratories to promote the binding of these vehicles to target cells by coupling specific ligands, so that the transported substances only become effective in the desired brain regions. Successes in this area of pharmaceutical research would be of enormous importance for the treatment of the majority of neurological diseases.

However, the results of clinical studies based on laboratory experiments are often difficult to predict. Ultimately, we lack the knowledge of how exactly altered proteins lead to neuronal degeneration. With regard to tau, for example, the following questions arise: Are the disease-causing forms of tau rather monomers, i.e. single incorrectly configured tau molecules? Or are they the tau oligomers or the fibrils? Do they develop their pathogenic effect intra- or only extracellularly? How important is it to stop the transfer of tau from neuron to neuron, i.e. to interrupt its prion properties? There is certainly a lot to do for basic research in the future.

## 3.2.7  Symptomatic Therapy

With regard to already existing and easily available therapies, there are promising data on a polyamine, the spermidine originally isolated from sperm. Its concentration clearly decreases with age in the body. Intriguingly, in animal models of neuronal degeneration, the administration of spermidine improves memory (probably through the stimulation of autophagy) and initial studies show that spermidine can delay the onset of dementia in older people. Other observations suggest that people who consume spermidine-rich foods such as whole grain products, yeast, fruit and vegetables live up to five years longer.

Interestingly, cholesterol-lowering drugs, in particular simvastatin, can lower levels of phosphorylated tau in cerebrospinal fluid. Coconut oil also seems to have a favorable effect, as it down-regulates the ADP-ribosylation factor 1 through its high content of medium-chain fatty acids and thus reduces the secretion and aggregation of Aβ. The same applies to ginkgo extracts, which may have positive effects through anti-oxidative and anti-inflammatory effects. However, the clinical efficacy in controlled studies is still lacking for all of these naturally occurring substances.

Furthermore, therapies for diabetes may also be used in Alzheimer's disease. In type 2 diabetes, muscle, liver and nerve cells respond less well to the hormone insulin from the pancreas. This in turn makes neurons less responsive to the neurotrophic effect of insulin (mediated by IGF receptors), which results in higher sensitivity to cellular stress. Therefore, it is likely that anti-diabetes drugs will be useful in the treatment and prevention of Alzheimer's disease probably also via indirect effects on the blood vessels.

In addition, the local inflammatory reaction in the area of senile plaques represents an important pharmacological target. As explained in the second chapter, activated microglia, the brain's own inflammatory cells, release cytokines such as interleukin-1β and tumor necrosis factor TNFα. These can activate the p38/MAP kinase signaling pathway in neurons and thus cause phosphorylation of tau proteins. Consequently, various inhibitors of this particular signaling pathway (e.g. neflamapimod) are in clinical trials to stop neuronal damage caused by inflammatory mediators. Clinical trials with other anti-inflammatory approaches, e.g. with azeliragon, an antagonist of a transmembrane receptor of the immunoglobulin superfamily (RAGE), or with pioglitazone, a medication for diabetes mellitus, which is also anti-inflammatory, were unfortunately not successful as yet. As mentioned above, the treatment of Alzheimer's patients with anti-inflammatory painkillers (e.g. with naproxen) has so far not shown convincing effects either. Anti-inflammatory drugs were tested in large studies in early and advanced stages of the disease, but also in pre-symptomatic patients who had no cognitive deficits at the beginning of taking the substance. In this context, it should be noted that frequent inflammations (e.g. tooth infections), adversely affect the course of neurodegenerative diseases. An intact immune system and corresponding body care (including our teeth) therefore help to prevent inflammatory diseases and thus to delay brain degeneration in general.

An interesting new development from the immunological field concerns the complement inhibitors. As mentioned above, the complement factors C1q and C3 play an important role in the elimination of synapses by microglia during development. Neuronal cells produce endogenous inhibitors of complement proteins (e.g. SRPX2) and can thus prevent the clearance of synapses. The stimulation of SRPX2 or the therapy with complement inhibitors (e.g. Crry-Ig) may possibly prevent the loss of synapses in Alzheimer's disease and perhaps also in other diseases of the brain after trauma or in schizophrenia, for example.

Finally, due to the changed neuronal excitability (see Fig. 2.10) electromagnetic methods may be suitable to regulate the network activities in deep brain regions. Induced gamma oscillations with a frequency of 40 Hz apparently lead to a reduction of Aβ plaques and phosphorylated tau after multimodal stimulation affecting different sensory pathways. This could explain the improved cognitive abilities in animal models of neuronal degeneration by electrical therapy. For transcranial stimulation with short ultrasound pulses (TPS) there is not yet sufficient evidence for efficacy, since placebo-controlled studies are still lacking.

## 3.2.8  Which Measures Promise the most Success?

In view of the difficulties of finding new therapies we should bear in mind in what way the risk of Alzheimer's disease can be reduced for each individual today. First of all, by controlling lifestyle factors (alcohol, smoking, blood pressure, fat and sugar levels, weight). Furthermore, our hearing must function, because people with hearing impairments are much more likely to become socially isolated, which is a significant risk factor for dementia. Exercise, preferably as group activity, also plays an important role due to its social component and practice of movement coordination. Various studies have shown that moderate physical activity (at least three times a week) leads to an improvement in declarative, episodic and spatial memory in a significant proportion of patients. Repeated cognitive training is also highly recommended. All of these measures are simple, easily available, cheap, have no side effects, but a positive impact on various civilization illnesses, too.

Stimulation of the brain by mental and physical activity does not prevent but it will clearly delay the onset of neuronal degeneration. We suspect that the activity-induced release of neurotrophic factors in the brain (in particular BDNF, NGF and IGF) improves the survival of neurons and promotes dendritic plasticity necessary for learning and memory in the hippocampus and in other brain areas as well. Furthermore, after physical exercise, the circulation is stimulated and inflammatory components of the disease are reduced.

In terms of regular physical activity, which is estimated to reduce the risk of Alzheimer's dementia by approximately 45%, a cellular mechanism was proposed in 2019 by which muscle movement could have a positive effect on the brain. During exercise the protein irisin is released, which is formed by cleavage from a membrane protein in muscle cells, the fibronectin type III domain-containing protein-5 (FNDC5), and promotes burning of fat as well as cardiovascular function. Since irisin is also transported through the blood-brain barrier, it is possible that it may stimulate neurogenesis and synaptic plasticity as well. The latter seems to be explained by an enhancement of the release of BDNF. Interestingly, soluble Aβ peptides lead to a significant decrease in irisin in Alzheimer's patients. Therefore, it is not surprising that the cognitive functions improve with activity, since the levels of irisin are normalized in animal experiments. Probably there is a number of other mediators acting between muscles and the brain, which may be released via extracellular vesicles (exosomes) and enter the circulation. This includes cathepsin B (CTSB), a cysteine protease, which, like irisin, may cross the blood-brain barrier and promote BDNF production.

**In a Nutshell**

- Acetylcholine esterase inhibitors have been approved for the treatment of Alzheimer's dementia. However, their effect is small and they cannot stop the progression of the disease.
- Secretase inhibitors that are supposed to prevent the increase of pathological Aβ peptides have failed in clinical studies. Alternatives substances that have fewer side effects are being sought.
- Neurotrophic factors, such as fibroblast growth factor, stimulate neurogenesis in the hippocampus, improve memory functions and inhibit the aggregation of Aβ and tau. However, the problem of application and bioavailability remains to be solved for neurotrophic factors and other proteins.
- In higher concentrations, antibodies against Aβ peptides slow down cognitive decline and reduce the amount of senile plaques. In addition, antibodies against pathological tau and inhibitors of tau aggregation are being tested.
- Novel substances that stabilize protein homeostasis and inhibitors of the cellular stress response are also under development.
- The control of lifestyle factors such as alcohol, smoking, blood pressure and weight, as well as regular physical and social activity, are currently the most effective measures to delay the onset and progress of dementia.

## 3.3    Diagnosis and Therapy of Neuronal Degeneration—quo vadis?

Finally, I would like to give an outlook on how research in the field of neurodegeneration and the development of new therapeutic approaches could continue. But before that, an important methodological aspect should be discussed, which is relevant especially for neuropathological examinations and the attempt to find the actual cause of a degenerative brain disease.

A direct connection between laboratory tests on the one hand and the underlying pathomechanism of the disease on the other hand is usually not possible (if the disease is not caused by monogenic factors). This is due in particular to the lack of a time course, i.e. the pathology cannot be made visible continuously at cellular or even molecular resolution in the same specimen with the methods available today. This is not possible, neither in humans nor in animals. Only then an accurate description of the pathomechanism underlying a given disease could be given. It is the old problem of the lack of causality despite known correlation, because until today we can only compare different brains and samples with each other in the required resolution to obtain meaningful data.

Therefore, it remains unclear in most cases whether an observed pathology is really the cause of a certain clinical symptomatology or perhaps a consequence of a completely different pathology which would not lead to the observed failures on its own. The microscopic or biochemical findings could therefore be a concomitant phenomenon without direct causal connection to the actual cause of the disease. The same applies to the typical laboratory data, mainly from blood or CSF, which are collected from a patient. This problem runs through practically all organs in medicine and is particularly relevant for many internal and neurological diseases.

Currently, we cannot predict whether there will ever be such high-resolution microscopical systems that allows us to analyze a specific disease process at the level of a single cell over time. Such a method would have to be non-invasive, as brain tissue is particularly sensitive and must not be tampered with in order to avoid additional pathology. In addition, the neuronal network context must not be disturbed. Only after the development of a completely new imaging technique (comparable to the invention of magnetic resonance imaging) would it be possible, in my opinion, to definitely clarify the pathomechanistic relationships discussed in this book for Alzheimer's or Parkinson's disease and to confirm them as causal at the cellular or molecular level.

Due to the complexity of the brain in general and of neurodegenerative diseases in particular, further progress in diagnosis and treatment will be necessary in order to spare future patients the disease or at least to delay its course. It is very unlikely that there will be an identical and causal therapy for all patients alike. The molecular mechanisms that lead to disturbances of endosomal transport, autophagy or mitochondrial energy production are too different to be remedied by a single targeted therapy. Especially in

Parkinson's disease, we also have to deal with neurons that play an indispensable role in large modulatory neuronal networks and are therefore very difficult to replace.

A key question will also be whether the deficits already present at the time of diagnosis can be positively influenced by any of the therapeutic approaches mentioned above. In recent years, hence, a shift in scientific interest to the asymptomatic (preclinical) stages of Parkinson's and Alzheimer's disease has taken place. Very early changes can be tested most easily in the inherited forms, since the relatives of a patient are still cognitively healthy, but carry an increased risk of developing the disease. New drugs will probably be used in the prodromal stage in the future, i.e. before the first symptoms set in. According to the current state of research, it can be assumed that neuropathological changes already occur in the brain about 20 years before the clinical manifestations become obvious.

Today, particular attention is already paid to disorders of the sense of smell and of the digestive system in order to be able to make an early diagnosis using advanced imaging techniques and laboratory diagnostics. This includes the discovery of new biomarkers for neurodegenerative diseases. For example, high uric acid levels in the blood correlate with a lower risk of Parkinson's disease. In contrast, an increase in phosphorylated tau proteins, in particular p181-tau, or defined tau fragments indicate the presence of a neurodegenerative process. Other neuronal proteins, e.g. the cytoskeletal neurofilament protein, are also discussed in this context. Initial studies show that the combined detection of p181-tau and the light chain of neurofilament (NFL) is a valuable prognostic factor for the onset of Alzheimer's dementia. After all, with this combination, more than 95% of the patients who will become demented within four years are pulled out with a single blood draw. On the other hand, this method has only low sensitivity, i.e. it will miss about half of the patients who will later become diagnosed with Alzheimer's disease.

The total amount of tau is elevated following brain damage in the cerebrospinal fluid. In Alzheimer's patients, total tau levels increase threefold. Such findings suggest a continuous rise of tau fibrils in the course of neurodegenerative diseases. It is assumed that in a few years from now a whole range of markers in cerebrospinal fluid and/or blood plasma will be available that can indicate the onset of a neurodegenerative disease.

If there is an early diagnosis but effective therapies are not yet available, a doctor would have to think carefully about whether to tell a clinically inconspicuous patient that he or she will soon be a Parkinson's or Alzheimer's patient. Because without the diagnosis, the person would probably feel much better for a few years than with the mental stress of knowing to be suffering from a chronic and ultimately fatal disease. An early diagnosis can be very depressing! As in a vicious circle, the reduced mood would then have negative effects on the still intact memory functions in the event of impending dementia. In situations like this, the attending physician has a special responsibility and needs to consider all diagnostic procedures very carefully.

Ultimately, we are reliant on further research performed in laboratories around the world that will show us a way to stop or slow down the process of neurodegeneration. This also requires new insights from comparative neurobiology (see first chapter). How has the organization of neuronal networks in the cortex of mammals changed over the millennia? Are there any living species today whose brains age more slowly and are less responsive to disturbances in protein homeostasis than the human brain? Perhaps evolution has produced other ways to regulate motor activities impaired in Parkinson's disease than the highly sensitive brainstem neurons with their complex axon morphology? A look into the CNS of other mammals may help us to discover fundamentally new approaches to how neurodegenerative diseases could be treated. Such research facilitates the *outside-the-box* thinking which enables us to take completely new directions in the therapy of particularly complex diseases.

In addition to the undoubtedly important genetic analyses, we should also pay more attention to the epigenetic changes, which, in contrast to DNA, can be modified throughout a lifetime. DNA-binding proteins that influence the formation (transcription) of mRNAs, but also the expression of long, non-coding RNAs, have come into the focus in many laboratories, because they—unlike the genes themselves—are subject to the influences of nutrition or stress throughout life. They can make our brain more or less sensitive to aging or prone to develop a neurodegenerative disease. This may happen over several generations, as epigenetic changes can also occur in germ cells and can therefore be passed on to our children.

Finally, a hopeful look into the future is appropriate. It is expected that neuroscientific research will continue to produce results at a very high speed. It has led us to a level of medical care that was completely unimaginable 100 years ago in the diagnosis and treatment of several diseases. The average life expectancy in industrialized societies is now about 90 years. This is not only due to a reduction in infant mortality, improved hygiene and the development of antibiotics. There are also a large number of drugs available now and great progress has been made in technical diagnostics (X-ray, MRI) and vaccinations (e.g. against polio, whooping cough, measles or Covid-19). High blood pressure and diabetes, but also epilepsy, can be treated well. This

**Fig. 3.5** A patient with dementia and her daughter (iStock.com/PIKSEL)

has not only increased the life expectancy, but also the quality of life in old age (Fig. 3.5).

If you liked this book and you want to find out more about the latest developments in the diagnosis and treatment of neurodegenerative diseases, I invite you to visit my website with a discussion of up-to-date articles on research and treatment of degeneration and regeneration in the nervous system (https://www.klimasbrainblog.com/en).

## Further Reading

Armstrong MJ, Okun MS (2020) Diagnosis and treatment of Parkinson disease: a review. JAMA 323:548–560

Aron L, Klein R (2011) Repairing the parkinsonian brain with neurotrophic factors. Trends Neurosci 34:88–100

Bennett ML, Bennett FC (2020) The influence of environment and origin on brain resident macrophages and implications for therapy. Nat Neurosci 23:157–166

Bijsterbosch J (2019) How old is your brain? Nat Neurosci 22:1611–1612

Connolly BS, Lang AE (2014) Pharmacological treatment of Parkinson disease: a review. JAMA 311:1670–1683

Deuschl G, Schade-Brittinger C, Krack P, Volkmann J et al (2006) A randomized trial of deep-brain stimulation for Parkinson's disease. N Engl J Med 355:896–908

Elia LP, Reisine T, Alijagic A, Finkbeiner S (2020) Approaches to develop therapeutics to treat frontotemporal dementia. Neuropharmacology 166:107948

Filipkowski RK, Kaczmarek L (2018) Severely impaired adult brain neurogenesis in cyclin D2 knock-out mice produces very limited phenotypic changes. Prog Neuro Psychopharmacol Biol Psychiatry 80:63–67

Hausott B, Klimaschewski L (2019) Sprouty2—a novel therapeutic target in the nervous system? Mol Neurobiol 56:3897–3903

Huang LK, Chao SP, Hu CJ (2020) Clinical trials of new drugs for Alzheimer disease. J Biomed Sci 27:18

Humpel C (2011) Identifying and validating biomarkers for Alzheimer's disease. Trends Biotechnol 29:26–32

Hwang JY, Aromolaran KA, Zukin RS (2017) The emerging field of epigenetics in neurodegeneration and neuroprotection. Nat Rev Neurosci 18:347–361

Jia Y, Nie K, Li J, Liang X, Zhang X (2016) Identification of therapeutic targets for Alzheimer's disease via differentially expressed gene and weighted gene co-expression network analyses. Mol Med Rep 14:4844–4848

Katsouri L, Ashraf A, Birch AM, Lee KKL, Mirzaei N, Sastre M (2015) Systemic administration of fibroblast growth factor-2 (FGF2) reduces BACE1 expression and amyloid pathology in APP23 mice. Neurobiol Aging 36:821–831

Kempermann G (2019) Environmental enrichment, new neurons and the neurobiology of individuality. Nat Rev Neurosci 20:235–245

Klimaschewski L, Claus P (2021) Fibroblast growth factor signalling in the diseased nervous system. Mol Neurobiol. 58:3884-3902

Lauzon MA, Daviau A, Marcos B, Faucheux N (2015a) Growth factor treatment to overcome Alzheimer's dysfunctional signaling. Cell Signal 27:1025–1038

Lauzon MA, Daviau A, Marcos B, Faucheux N (2015b) Nanoparticle-mediated growth factor delivery systems: a new way to treat Alzheimer's disease. J Control Release 206:187–205

Leavitt BR, Tabrizi SJ (2020) Antisense oligonucleotides for neurodegeneration. Science 367:1428–1429

Liu B, Teschemacher AG, Kasparov S (2017) Astroglia as a cellular target for neuroprotection and treatment of neuro-psychiatric disorders. Glia 65:1205–1226

Liu Z, Chopp M (2015) Astrocytes, therapeutic targets for neuroprotection and neurorestoration in ischemic stroke. Prog Neurobiol 144:103–120

Miklas JW, Brunet A (2020) Support cells in the brain promote longevity. Science 367:365–366

Moreno-Jiménez EP, Flor-García M, Terreros-Roncal J, Rábano A, Cafini F, Pallas-Bazarra N, Ávila J, Llorens-Martín M (2019) Adult hippocampal neurogenesis is abundant in neurologically healthy subjects and drops sharply in patients with Alzheimer's disease. Nat Med 25:554–560

Otsuki L, Brand AH (2020) Quiescent neural stem cells for brain repair and regeneration: lessons from model systems. Trends Neurosci 43:213–226

Palasz E, Niewiadomski W, Gasiorowska A, Wysocka A, Stepniewska A, Niewiadomska G (2019) Exercise-induced neuroprotection and recovery of motor function in animal models of Parkinson's disease. Front Neurol 10:1143

Panza F, Lozupone M, Logroscino G, Imbimbo BP (2019a) A critical appraisal of amyloid-β-targeting therapies for Alzheimer disease. Nat Rev Neurol 15:73–88

Panza F, Lozupone M, Seripa D, Imbimbo BP (2019b) Amyloid-β immunotherapy for Alzheimer disease: is it now a long shot? Ann Neurol 85:303–315

Parmar M, Grealish S, Henchcliffe C (2020) The future of stem cell therapies for Parkinson disease. Nat Rev Neurosci 21:103–115

Pena-Diaz S, Pujols J, Ventura S (2020) Small molecules to prevent the neurodegeneration caused by alpha-synuclein aggregation. Neural Regen Res 15:2260–2261

Pfisterer U, Khodosevich K (2017) Neuronal survival in the brain: neuron type-specific mechanisms. Cell Death Dis 8:e2643

Pramanik S, Sulistio YA, Heese K (2017) Neurotrophin signaling and stem cells—implications for neurodegenerative diseases and stem cell therapy. Mol Neurobiol 54:7401–7459

Salamon A, Zádori D, Szpisjak L, Klivényi P, Vécsei L (2019) Neuroprotection in Parkinson's disease: facts and hopes. J Neural Transm 127:821–829

Sell GL, McAllister AK (2020) Protecting connections from synapse elimination. Trends Neurosci 43:841–842

Simrén J, Ashton NJ, Blennow K, Zetterberg H (2020) An update on fluid biomarkers for neurodegenerative diseases: recent success and challenges ahead. Curr Opin Neurobiol 61:29–39

Sun J, Roy S (2021) Gene-based therapies for neurodegenerative diseases. Nat Neurosci 24:297–311

Titze de Almeida SS, Soto-Sánchez C, Fernandez E et al (2020) The promise and challenges of developing miRNA-based therapeutics for Parkinson's disease. Cell 9:841

Yamashita N, Kuruvilla R (2016) Neurotrophin signaling endosomes: biogenesis, regulation, and functions. Curr Opin Neurobiol 39:139–145

Zhao HT, John N, Delic V, Ikeda-Lee K, Kim A, Weihofen A, Swayze EE, Kordasiewicz HB, West AB, Volpicelli-Daley LA (2017) LRRK2 antisense oligonucleotides ameliorate α-Synuclein inclusion formation in a Parkinson's disease mouse model. Mol Therapy Nucl Acids 8:508–519

# Glossary

α-**Synuclein** This 140 amino acid long protein is highly relevant for Parkinson's disease. It is involved in cellular transport processes and can form holes in the plasma membrane.

**Acetylcholine** One of the most important transmitters of the peripheral and central nervous system, which was first demonstrated by Otto Loewi (1921) in the frog heart. It is an ammonium compound as an ester of acetic acid and the amino alcohol choline.

**Allocortex** In addition to the neocortex (also called isocortex), there is the 'other' cortex (Greek, allos), which does not have the typical six-layered structure of the neocortex (see Cortex cerebri).

**Amygdala** see Corpus amygdaloideum

β-**Amyloid** The cleavage products of the amyloid precursor protein (APP) include the Aβ peptides (e.g. Aβ42, Aβ40). They normally have antimicrobial functions, but at higher concentrations they can damage the brain because they have a tendency to aggregate and form plaques. The β-amyloid is the main component of amyloid plaques in Alzheimer's disease.

**Amyloid-Precursor-Protein (APP)** An integral membrane protein of over 700 amino acids that is cleaved by the secretases.

**Anosmia** The loss of the sense of smell is called anosmia, a reduction in smell is called hyposmia. However, certain stimuli such as ammonia are still perceived, as they activate the trigeminal nerve and are conveyed via this nerve (and not via the olfactory nerve) to the brainstem.

**Antisense oligonucleotides (ASOs)** These are artificially produced, short-chain single strands of nucleic acids. Since they are opposite in base sequence to the mRNA, they can hybridize with the mRNA (hence *anti-sense*). They block the translation, i.e. the production of the protein encoded by the mRNA at the ribosome.

L. P. Klimaschewski, *Parkinson's and Alzheimer's Today*,
https://doi.org/10.1007/978-3-662-66369-1

**Aphasia** If the understanding of spoken or written language is intact, but speech is impaired and word-finding difficulties occur, this is referred to as motor aphasia. It is usually associated with damage to the area named after Broca in the inferior frontal gyrus of the frontal lobe. In pure sensory aphasia, the understanding of language is restricted only. Those affected are not able to grasp the meaning of sounds and their understanding of music is reduced. There may be word substitutions (semantic paraphasias) and word creations (neologisms). Often, this form of language disorder affects the area named after Wernicke in the superior temporal gyrus.

**Apoptosis** Apoptosis (Greek, falling off) refers to programmed cell death, i.e. the activation of a suicide program in cells. In contrast to necrosis, the cell dies without damage to the surrounding tissue. Apoptosis can be induced from the outside, e.g. by immune cells, or started intrinsically by DNA damage or cellular stress.

**Aquaeductus mesencephali** As a narrow connection between the third and fourth ventricles in the midbrain, the aqueduct separates the posterior roof (tectum) from the area in front of it (tegmentum). In the tegmentum, in addition to the red nucleus, the substantia nigra is located.

**Association tracts** This term refers to all fiber bundles that connect different cortical areas of one hemisphere with each other. They therefore do not cross to the opposite side as the commissural tracts do.

**Astrocytes** see Glia

**Autophagy** A cellular process that degrades cell-internal structures (misfolded proteins, protein aggregates, organelles) and recycles the components.

**Axon** The singular, often also longest extension of a nerve cell is called an axon or neurite or nerve fiber, if it includes a myelin sheath. Lateral branches of the axon (also referred to as axon collaterals or axon branches) allow the establishment of a connection with various other neurons (divergence). In the area of the axonal endings (terminals) there are usually several synapses that make contact with the target cells in the form of a "presynapse". Presynapses are located exactly opposite to the postsynaptic specializations on the dendrites of innervated nerve cells. Stacked microtubules give the axon strength and allow the transport of intracellular vesicles (endosomes) and mitochondria.

**Basal ganglia** The basal nuclei are located below the cortex and are arranged around the lateral ventricles and the diencephalon. The essential parts of the basal ganglia arise from the ganglionic eminence formed in the basal part of the telencephalic hemisphere. The "tail" nucleus (nucleus caudatus) and the putamen (Latin for shell) are together referred to as the striatum (corpus striatum). In addition, the structures arising from the diencephalon and mesencephalon, the globus pallidus (Latin for pale sphere) and the substantia nigra (Latin for black substance), are also considered part of the basal ganglia. Their neurons are primarily involved in motor circuits, i.e. they regulate the motor and premotor areas of the cortex. The basal ganglia control our movements and behavior, i.e. all somato- and psychomotor processes. The nucleus accumbens, located in the lower, anterior striatum, plays an important role in the latter (see below).

**Brainstem** see Truncus cerebri

**Brodmann areas** The cortical areas named after Korbinian Brodmann (1868–1918), which he distinguished and numbered on the basis of histological and cytoarchitectural features. The Nissl staining of nerve cells developed by Franz Nissl during his medical studies allowed Brodmann to describe the areas in detail. For example, the primary visual cortex is referred to as Area 17 (Area striata). In Brodmann's Areas 1–3, we consciously perceive the sensations emanating from our body surface (somatosensory cortex). In front of it is the primary motor cortex as part of the precentral gyrus (Area 4).

**Bulbus olfactorius** The olfactory bulb is part of the telencephalon (forebrain). Its cytoarchitecture has several layers of cells that process signals from the olfactory epithelium in the upper nasal cavity. The results of these calculations are forwarded via the olfactory tract to the paleocortex. Neurogenesis takes place here throughout life.

**Capsula interna** The internal capsule encloses most of the ascending and descending tracts of the cortex cerebri. In a horizontal section through the brain, it is V-shaped and is located between the thalamus (diencephalon) and the basal ganglia. The caudate nucleus and the putamen (shell) are separated by the internal capsule.

**Caspases** Caspases are cysteine proteases that cut their target proteins next to the amino acid aspartate (hence the name). They are important enzymes involved in neuronal apoptosis.

**Catecholamines** The messenger substances characterized by a common amino group, the catecholamines, include dopamine, norepinephrine, and epinephrine. They are formed from the amino acid tyrosine with the help of the enzyme tyrosine hydroxylase, have hormonal functions, and bind to G protein-coupled receptors (adrenoceptors or dopamine receptors).

**Cerebellum** The cerebellum lies dorsally to the brainstem. It is a separate part of our brain that develops at the roof of the fourth ventricle and has a centrally located worm (vermis) and two hemispheres. The cerebellum is involved in the fine-tuning and coordination of motor programs, but also in cognitive processes. It is of great importance for maintaining balance and making quick corrective movements. In humans, it accounts for only about one tenth of the brain weight with about 150 g, but contains up to 70 billion neurons and thus at least 80% of all nerve cells in the human central nervous system. On the surface of the cerebellum are the folia and fissures (leaf-shaped elevations and depressions), instead of gyri and sulci which are forming the cortex cerebri. The cortex of the folia is three-layered and much simpler in structure than the cortex of the gyri in the cerebrum. In the medullary cavity of the cerebellum are four cerebellar nuclei, including the largest dentate nucleus.

**Cerebrospinal fluid** The fluid located in the brain and in the spinal cord is called cerebrospinal fluid (CSF). It is produced by differentiated epithelial cells of the choroid plexus located in the ventricles. It is a clear, colorless solution and contains some protein, sugar and only a few cells (lymphocytes). The cerebrospinal fluid is drained from the arachnoid villi of the dura mater into the venous sinuses (sinus durae matris) and into the lymphatic space beyond the exit points of cranial and spinal nerves.

**Cerebrum** The cerebrum is also called telencephalon (Greek). It forms the two large hemispheres with their 4 lobes, all of which are covered with cortex. Underneath are the basal ganglia, which have arisen from the ganglionic eminence. The first cranial nerve, the olfactory nerve, is an extension of the cerebrum.

**Chaperone** Chosen after the English term for 'chaperone' or 'nurse' it describes a protein that helps another protein to fold, i.e. to achieve the correct three-dimensional secondary and tertiary structure after its amino acid chain (the primary structure) has been generated at the ribosome. Chaperones are of particular importance for proteins that are prone to aggregation.

**Commissure** It is a connection between the hemispheres, i.e. a fiber bundle crossing the median plane (midline). Connected areas usually correspond to each other structurally and functionally.

**Corpus amygdaloideum** The amygdala (almond-shaped nucleus) is a complex of several smaller nuclei located in the rostro-medial (i.e. anterior-medial) part of the temporal lobe. The amygdala plays a central role in the emotional evaluation and recognition of dangerous situations. Direct connections to the hypothalamus, which controls our vegetative nervous system, lead to the feelings associated with fear and anxiety. However, in general, all affects, i.e. also pleasurable sensations, are linked to an intact amygdala.

**Corpus callosum** The corpus callosum, with its approximately 250 million axons in humans, is the largest left-right connection (commissure) in our brain. Cortex regions located on both sides at a comparable level are interconnected via the corpus callosum. It is located in the depths of the longitudinal fissure between the two hemispheres (fissura longitudinalis) and thus also forms the roof of the two lateral ventricles.

**Corpus mamillare** Paired elevation at the base of the brain between the cerebral crura of the midbrain. In the corpora mamillaria, numerous fibers of the fornix end and therefore belong to the limbic system. The nerve cells of the mammillary bodies form essential nuclei of the posterior hypothalamus (diencephalon).

**Corpus striatum** The striatum is the largest nucleus of the basal ganglia. It consists of the caudate nucleus and the putamen, which are separated from each other by the ingrowing fibers of the internal capsule during embryonic development. Some groups of neurons still present in the area of the fiber masses of the inner capsule give the structure a striped appearance (therefore the name, striatum). The neocortex activates the corpus striatum in the sense of a one-way street, i.e. there are no axonal projections from the striatum back to the cortex. The cortico-striatal connections play a central role in the interaction of motivation and cognition with the voluntary and involuntary motor activities. These can originate from all lobes of the cortex, passed on to the striatum and further to the anterior and lateral ventral thalamic nuclei from which various motor areas of the cortex are activated. They directly innervate neurons in the brainstem and spinal cord, which ultimately cause our muscles to contract. In addition to these somatomotor functions, the nucleus accumbens, which is important for psychomotoric behavior, is located in the anterior and phylogenetically older part of the striatum.

**Cortex cerebri** In the cortex of the telencephalic hemispheres, three types of cortex are distinguished on the basis of histological features: 1. The paleocortex is the phylogenetically oldest part located in the lower (ventral) frontal and anterior temporal lobes. It contains 1-2 layers of neurons. 2. The archicortex is phylogenetically somewhat younger than the paleocortex and three-layered. It is located on the inside of the temporal lobe and consists of the hippocampus, the gyrus dentatus and the structures associated with them. These represent the core areas of the limbic system. 3. The neocortex occupies 90% of the entire cortex and is the youngest, six-layered part. According to Brodmann, many more cortical subtypes can be distinguished due to their special cytoarchitecture. For example, individual layers of the cortex are not present in all neocortex areas, and certain cell types are more common in some areas than in others.

**Cortex entorhinalis** The entorhinal cortex (Brodmann areas 28 and 34) covers the rostral (anterior) part of the parahippocampal gyrus. This area located at the medial (inner) edge of the temporal lobe marks the transition from the allocortex to the neocortex. From here, the cells are already arranged in six layers and project into the hippocampus, in particular to the dentate gyrus. This connection is called the perforant path (tractus perforans). In the entorhinal area, excitations from the olfactory, somatosensory, visual and auditory sensory centers converge. This makes the entorhinal cortex a multimodal association center, which is involved in storing and retrieving declarative (semantic and episodic) memory content in connection with the hippocampal formation.

**Cortex insularis** The insula (island) is also called the fifth brain lobe and is phylogenetically old. In prenatal development, its growth is slower than that of other lobes, so that the insula is covered by the other lobes. This creates a depression on the outside of the hemisphere, the sulcus lateralis. The insular cortex belongs to the neocortex and is also referred to as the paralimbic cortex due to its connections to limbic structures. The posterior part of the insula processes pain and other sensations as well as acoustic information. Anterior areas are primarily responsible for olfactory and gustatory sensations and general visceral information from the intestine.

**Cortex orbitofrontalis** The orbitofrontal cortex is a part of the prefrontal cortex and is located above the orbita (that is the eye socket). It connects our emotions, anchored in the limbic system, and rational-cognitive functions processed in the dorsolateral prefrontal cortex. This is due to its extensive connections with the amygdala and the mediodorsal thalamic nucleus. The orbitofrontal cortical areas constantly evaluate the emotional stimuli coming from the deeper brain regions. Injuries to these areas therefore lead to severe changes in personality (emotional impoverishment, indifference, but also disinhibition), and the neurotransmitter acetylcholine plays an important role here.

**Cortex prefrontalis** The primary function of the prefrontal cortex areas is monitoring. Various connections to other cortical regions and nuclei enable the analysis and evaluation of information. The results are then sent back to the respective brain areas. The function of the dorsolateral cortex is thus similar to the central processor (CPU) in our computers. Damage to the frontal lobe results in reduced cognitive flexibility, i.e. the person can no longer adjust to new situations. Insignificant stimuli are no longer distinguished from relevant ones. However, everyday routines (shopping, preparing meals, etc.) usually proceed without interruption.

**Cytokines** These are small proteins that regulate the growth and differentiation of cells. On the one hand, they behave like growth factors, on the other hand they act as mediators of immune reactions and inflammatory processes (e.g. interleukins and the tumor necrosis factor, TNF-α).

**Cytoskeleton** see Microtubule

**Dendrite** The cellular extensions of nerve cells that receive stimuli are called dendrites. In addition to the cell body (soma or perikaryon), a neuron usually has numerous dendrites and a single axon. Dendrites have spines, which carry the post-synaptic contact site, i.e. they bind neurotransmitters released from the pre-synapse.

**Diencephalon** This hidden part of the forebrain follows the midbrain cranially and encloses the third ventricle. The largest diencephalic nucleus is the (dorsal) thalamus, also called the "secretary of the boss", i.e. all sensory sensations with the exception of the sense of smell have to pass through the thalamus in order to reach the neocortex (the "boss"). Below the dorsal thalamus are the subthalamus and the hypothalamus, above the epithalamus. The hypothalamus is the vegetative center for the control of metabolism, blood pressure, heat and water balance, sweat secretion and genital functions. The inner part of the pallidum belongs to the basal ganglia but is a derivative of the diencephalon and, together with the subthalamus, controls motor functions.

**Dopamine** A biogenic amine whose name is composed of DOPA and amine. It is a key neurotransmitter in the central nervous system and belongs to the catecholamines. Dopamine is the immediate precursor of melanin, which gives the dopamine and noradrenaline producing brainstem nuclei a bluish-black color.

**Embryogenesis** This is the earliest phase of our development, i.e. from the fertilized egg (zygote) via various intermediate stages (blastulation, gastrulation, neurulation) to the formation of organs (organogenesis).

**Endoplasmic Reticulum (ER)** The ER is a widely branched, plasma membrane-enclosed network of tubules and cavities. The ER membrane directly transitions into the nuclear envelope of the cellular nucleus. Parts of the ER are occupied by ribosomes for the production of proteins that are synthesized directly into the lumen of the ER or into the ER membrane. It is therefore referred to as "rough" ER or ergastoplasm (in contrast to the ribosome-free "smooth" ER). In addition to translation, the ER also serves as a site for protein folding, quality control of newly formed proteins, certain modifications of proteins, protein transport and synthesis of lipids. In addition, the ER acts as intracellular calcium store.

**Endocytosis** Fluid, molecules, and particles are taken into the interior of the cell by invaginations of outer or inner membranes forming vesicles (endosomes).

**Endosomes** Also known as endosomal vesicles, these cell organelles are primarily formed by endocytosis. Early and late endosomes are distinguished. Various membrane proteins are transported via late endosomes into lysosomes where they are degraded or recycled back to the cell membrane via recycling endosomes.

**Encephalisation quotient (EQ)** This value sets the individually measured brain weight in relation to that brain weight which would be expected for a given species with comparable body weight.

**ERAD complex** Faulty folded or mutated proteins that enter the endoplasmic reticulum (ER) can be secreted back into the cytoplasm from the ER and, after ubiquitination, degraded outside the ER by the proteasome. The enzymes involved in this process are grouped together in the ERAD complex (**ER-a**ssociated **D**egradation) and also referred to as the unfolded protein response (UPR) or, in the case of overload, as ER stress.

**Evolution** Evolutionary biology deals with the inheritable traits of a population of living beings that are passed on from generation to generation. In the broadest sense, this also includes the development of bacteria and viruses over time.

**Exocytosis** In contrast to endocytosis, this refers to a transport of substances out of the cell. Vesicles from the cytoplasm fuse with the cell membrane and, in doing so, they release the substances stored in them into the extracellular space.

**Fornix** This is an axonal fiber bundle that runs under the corpus callosum and connects the hippocampus with the hypothalamus, in particular with the corpora mamillaria. Part of the fibers pass over the commissura fornicis to the opposite side, thus establishing an important connection between the two hippocampal structures.

**GABA** As the most important inhibitory neurotransmitter, γ-aminobutyric acid (GABA) is formed by decarboxylation of glutamic acid in neurons.

**Genome-wide association studies (GWAS)** In the search for a genetic causes of disease, extensive DNA sequencing is carried out. This makes it possible to identify alleles (variants of a gene) that occur together with a certain phenotype (trait). In particular, genetic markers (SNPs, single nucleotide polymorphisms) are used to find conspicuous DNA sections that are usually not located in a protein-coding region, but rather in non-coding regions between two coding genes. Due to the ever-decreasing costs for DNA sequencing and polymerase chain reactions (PCR), GWAS are very popular for discovering associations, but not necessarily causal relationships, between our genome and a disease.

**Glia** All cells of the nervous tissue that can be distinguished structurally and functionally from nerve cells (neurons) are referred to as glial cells. This includes in particular the astrocytes (star cells) and the oligodendrocytes (together called macroglia) as well as the immunocompetent microglial cells, which perform monitoring and cleanup tasks. Furthermore, ependymal cells, which line the brain cavities (ventricles), and the epithelial cells of the choroid plexus that produce cerebrospinal fluid are also considered glial cells. Glia was first described in the 19th century by Rudolf Virchow (1821–1902) and referred to as glue (Greek, glia), which holds the neurons together. For Virchow, the supporting and holding function of the cells was in the foreground. He could not know that the glia is also involved in substance exchange and repair processes.

**Globus pallidus** The pallidum is an important part of the basal ganglia. The inner (medial) part is a derivative of the diencephalon, the outer (lateral) part arises from the ganglionic eminence of the telencephalon. Important afferent (incoming) pathways originate in the striatum and the nucleus subthalamicus. The most important efferents (outputs) project to the ventral and anterolateral nuclei of the thalamus via a pathway called ansa lenticularis. Functionally, the pallidum inhibits thalamic neurons, so that motor programs are overall activated due to double inhibition (its input from the striatum is also inhibitory). A failure of the globus pallidus therefore results in a lack of movement and clumsiness.

**Glutamate** Glutamic acid is a protein-forming amino acid and the most important excitatory neurotransmitter in the central nervous system.

**Glymphatic system** A neologism from the terms "glia" and "lymphatic system", which is thought to serve, similar to the lymph of the body, in the disposal of proteins, particles and other waste products in the brain. The transported substances then reach the veins and lymph with the cerebrospinal fluid (CSF). The glymphatic, perivascular space is located between astrocytic processes and the vessel wall (Virchow-Robin space).

**Golgi apparatus** This refers to the stack of intracellular membranes named after the Italian pathologist Camillo Golgi at the end of the 19th century, which is involved in secretion and other tasks of cellular metabolism. It is usually located near the cell nucleus and polarized: one side is convex, faces the endoplasmic reticulum (ER), and receives pinched-off vesicles (cis-Golgi network). The concave side facing away from the ER is called the trans-Golgi network (TGN). From there, secretory vesicles reach the outer plasma membrane and release their contents by exocytosis. The Golgi apparatus therefore forms a network of several stacked cisterns and vesicles that are closely interconnected.

**Granular neurons** The smaller neurons, also known as granular cells, have a round cell body (perikaryon). They are mostly interneurons with shorter axons that contact adjacent nerve cells. In the neocortex, they are found primarily in the second and fourth layer (stratum granulosum externum et internum).

**Growth factors** These proteins, for example, the members of the neurotrophin or fibroblast growth factor (FGF) families, have a signaling function, i.e. they serve to transmit information between cells in an organ (similar to hormones in the blood). They therefore play a particularly important role during the development of multicellular organisms, but also in the maintenance and repair of mature organs. Signal transmission usually takes place via the binding of the growth factor to a specific receptor located in the cell membrane of the target cell. This can also be the cell that produces the factor itself (i.e. an autocrine effect).

**Gyrus cerebri (plural Gyri)** Gyri are the convolutions between the brain furrows (sulci). They form a typical surface relief that is different for each person and are covered by the six-layered cortex cerebri.

**Gyrus cinguli** As an important part of the outer limbic gyrus, the cingulate gyrus surrounds the corpus callosum. Below its cortex, the cingulum, a long association tract, runs through the white matter, thereby establishing numerous connections between the frontal, temporal and parietal lobes. Close contacts with the anterior thalamic nucleus and also with the hypothalamus have been described. The anterior cingulate gyrus is activated when we perceive the suffering of other people and respond empathically by changing our behavior so that the suffering of the affected person is reduced. Evolutionarily, this area is therefore considered a center for morality and altruism. Lesions of the cingulate gyrus lead to the loss of drive, indifference and an emotionless, dull behavior.

**Gyrus dentatus** The so-called fascia dentata is a part of the hippocampal formation. The afferents arise from the entorhinal cortex and reach the dendrites of the small neurons (granule cells) of the gyrus dentatus via the tractus perforans (perforant path). Connections exist with the hypothalamus, the septal nuclei and the cingulate gyrus.

**Hemisphere** The respective, largely symmetrically constructed halves of the cerebrum or cerebellum are referred to as hemispheres.

**Hippocampus** The hippocampus is a brain structure located at the floor of the lateral ventricle in the temporal lobe and named after the seahorse. This is modeled after the sea monster Hippocampus from Greek mythology, whose front half is a horse and whose rear part is a fish (from hippos, horse and kampos, sea monster). It forms the central structure of the inner limbic loop and is made up of a three-layered archicortex. Here, information from various sensory systems are processed and then sent back to the neocortex. The hippocampus therefore plays a central role in memory consolidation, i.e. the transfer of memory content from short- to long-term memory. Patients who have lost both hippocampi can not form new memories. They have an "anterograde" memory loss (amnesia) starting from the time of degeneration. Older memories usually remain stable. Long-term storage of memory content does not take place in the hippocampus, but in the neocortex. Particularly important content is stored in several locations. In addition, the hippocampus plays a central role in our emotions. In chronic emotional stress, nerve cells in the hippocampus may be lost and new ones can not be produced (endogenous neurogenesis stops in the adult nervous system). In patients with severe mental trauma (e.g. in war or rape victims), a reduction in hippocampal volume can be demonstrated.

**Hypophysis** The pituitary gland (glandula pituitaria) consists of the neurohypophysis (posterior pituitary), derived from the diencephalon, and the adenohypophysis (anterior pituitary) arising from the roof of the mouth (Rathke's pouch). The hyophysis releases a plethora of hormones that regulate various processes such as growth, reproduction, and metabolism via direct effects or via activation of endocrine organs.

**Hypothalamus** The hypothalamus is the center of our vegetative, autonomously working nervous system, and mainly acts by releasing hormones affecting the glandular cells of the adenohypophysis. It is closely connected to limbic structures and controls respiration, circulation, body temperature, digestion, fluid balance, sexual function, and body growth during development.

**Insula** see Cortex insulae

**L-Dopa** L-Dopa is a precursor in the production of adrenaline, noradrenaline and dopamine from the amino acid tyrosine.

**Lewy body** The neuronal inclusion bodies first associated with Parkinson's disease by Friedrich Lewy (1885–1950) are also detectable in Lewy body dementia. In particular, α-synuclein, ubiquitin and neurofilaments are found in these mixtures of protein aggregates and cellular organelles.

**Limbic system** The limbic system plays a decisive role in emotionality, memory formation, and motivation. Its components form a double ring around the basal ganglia and the thalamus, respectively. It consists mainly of phylogenetically older cortical areas (paleocortex and archicortex) and subcortical structures that are mainly located in the temporal lobe. The outer limbic arc is represented by the cingulate and the parahippocampal gyrus, the inner arc comprises the fornix and the hippocampus. In addition, the amygdala (corpus amygdaloideum), the anterior thalamic nucleus and the mammillary bodies (corpora mamillaria) are included, too.

**Locus coeruleus** The "blue spot" belongs to the aminergic cell groups that are located at the floor of the fourth ventricle. It contains about half of all noradrenergic cells. The dark blue pigmentation results from an accumulation of neuromelanin, a product of catecholamine metabolism. Activation of the locus coeruleus results in the release of noradrenaline as a transmitter in various regions of the brain and the spinal cord.

**Long-term potentiation (LTP)** A form of morphological and biochemical neuroplasticity that takes place exclusively at synapses. It is a long-lasting (long-term) enhancement of synaptic transmission, so that the synaptic contacts have an increased probability of generating an action potential in the target cell. LTP is the basis for most learning processes in the brain.

**Lysosome** Cell organelles enclosed by a plasma membrane that have an acidic pH (4-5). Their essential function is the intracellular digestion of material by hydrolyzing enzymes such as proteases, nucleases, and lipases.

**Magnetic resonance imaging** Magnetic resonance imaging (MRI) is used to generate images that represent the structure and function of body organs. The technique is based on strong magnetic fields that excite atomic nuclei (usually hydrogen nuclei, protons) in the body. This results in an electrical signal that is detected by MRI.

**Medulla oblongata** The medulla oblongata (myelencephalon) is the lower part of the brainstem and about 3 cm long. Together with the pons and the cerebellum, the medulla oblongata forms the rhombencephalon, which surrounds the fourth ventricle. The medulla oblongata contains the vital centers for the regulation of circulation and respiration. These are mainly controlled by the hypothalamus. In addition, several reflexes necessary for the survival of the organism are neuronally interconnected in the medulla oblongata (sucking, swallowing, coughing, sneezing, vomiting).

**Medulla spinalis** The spinal cord is the part of the central nervous system (CNS) located in the spinal canal and, like the brain, is surrounded by membranes (meninges). It establishes connections with the peripheral nervous system via the spinal nerves. The outer parts (white matter) contain the ascending and descending fiber tracts (bundles of axons), while inside, surrounding the spinal canal, defined columns of neurons (gray matter) are located for the transmission of motor and sensory impulses. In the anterior horn of the gray matter the somatomotor neurons are situated, in the posterior horn the sensory afferents from the periphery enter the grey matter and become synaptically connected. We distinguish 8 cervical, 12 thoracic, 12 lumbar and 5 sacral segments of the spinal cord.

**Microglia** see Glia

**Microtubules** These are tubular, large protein complexes that, together with actin filaments and intermediate filaments, make up the cytoskeleton. They stabilize the shape and structure of the cell and are necessary for intracellular transport processes and active movements of whole cells or their cytoplasmic processes.

**Mitochondria** In contrast to lysosomes or the endoplasmic reticulum (ER), mitochondria are enclosed by a double plasma membrane and contain their own genetic material (mitochondrial DNA). They occur as spherical or tubular organelles and represent the cell's power plants by producing the energy-rich molecule adenosine triphosphate (ATP) via the respiratory chain. This process is also known as oxidative phosphorylation. Mitochondria multiply by growth and budding, depending on the energy needs of the cell.

**Midbrain (Mesencephalon)** The part of the brainstem that lies between the pons and the diencephalon. In fish and reptiles, the midbrain largely takes over the functions of our forebrain, i.e. the afferents of the visual and auditory pathways as well as of surface sensations end here. The midbrain controls most of the eye muscles and contains the substantia nigra and the nucleus ruber (red nucleus), i.e. essential components of the extrapyramidal motor system.

**Monoamine oxidases** Mitochondrial enzymes that break down monoamines (serotonin, dopamine, norepinephrine, and epinephrine) by deamination to aldehydes, ammonia, and hydrogen peroxide.

**Motor cortex** The gyri also known as the motor cortex are those cortical areas in the frontal lobe of our brain from which voluntary movements are initiated. We distinguish primary motor, premotor and supplementary motor areas (from back to front). The projection neurons (pyramidal cells) in lamina V of the motor cortex have very long axons (down to the lower spinal cord) and therefore relatively large cell bodies with pyramidal shape.

**Multivesicular bodies (MVBs)** These organelles are formed from early endosomes by multiple invaginations of their surrounding membrane. There are numerous smaller, membrane-enclosed vesicles inside the MVB, mostly of round shape. A fusion of MVBs with lysosomes leads to the breakdown of all their contents.

**Myelin** Myelin, discovered in 1854 by Rudolf Virchow, is formed by oligodendrocytes in the central nervous system and surrounds the axons of myelinated neurons. The spiral-shaped myelin sheath accelerates the conduction velocity by allowing electrical charges on the plasma membrane (the action potentials) to "jump" from one narrow, unmyelinated area between two glial cells to the next (a process called saltatory conduction). Since myelin consists of stacked plasma membranes, it has a high lipid content (70%, the rest are proteins) and appears white to the eye (white matter of the brain).

**Neocortex** see Cortex cerebri

**Nervus opticus** The optic nerve is the second cranial nerve. It belongs to the central nervous system because it is derived from the diencephalon. The optic nerve carries the axons of the retinal ganglion cells of the eye to the optic chiasm, which is located above the pituitary gland. In the optic chiasm, the fibers of the nasal halves of the retina cross to the opposite side, so that the left visual field is mapped in the right occipital lobe (and vice versa).

**Nervus vagus** The tenth cranial nerve is the largest nerve that branches widely in the chest and abdominal cavity (Latin vagare, to wander). It belongs to the autonomic (parasympathetic) nervous system and regulates almost all organ activities with efferent (motor) and afferent (sensory) nerve fibers. Its motor neurons are located in the brainstem (medulla oblongata, nucleus dorsalis n. vagi) and below the base of the skull in sensory ganglia (nodose and jugular ganglion).

**Neurogenesis** The formation of nerve cells from dividing stem cells is called neurogenesis. It takes place in humans during development up to the early postnatal phase. Hippocampal neurogenesis is observed until the age of 20. Thereafter, it is hardly detected outside the olfactory bulb. However, individual remaining stem cells could possibly be stimulated to divide and generate new neurons in the adult brain.

**Neurotrophins** Signaling substances in the nervous system that play a key role in the formation of connections between nerve cells (with each other and with their peripheral effectors, such as muscles, glands, or the skin) mainly during development. They are small basic proteins with a molecular weight of appr. 13 kDa that are primarily responsible for the maintenance of neurons and the outgrowth of their processes (dendrites and axons).

**Noradrenaline** A biogenic amine that acts as a stress hormone and neurotransmitter. Like adrenaline, it leads to the constriction of blood vessels and increases in blood pressure. In the brain, it is produced primarily by the locus coeruleus.

**Nuclei raphes** The raphe nuclei form several groups of neurons located along the midline (Greek, seam) and produce serotonine. They innervate the cortex and the spinal cord, but also the basal ganglia. In the spinal cord, they play an important role in endogenous pain inhibition by activating interneurons that control nociceptive fibers from peripheral sensory neurons.

**Nuclei septi** The septum is located between the lateral ventricles. The nuclei located in the lower part of the septum are involved in complex functional neuronal circuits between the diencephalon (hypothalamus) and the temporal lobe. They are therefore part of the limbic system. Via the fornix, they are reciprocally connected to the hippocampus and via the stria terminalis to the amygdala.

**Nucleus accumbens** In the frontal and ventral (caudal) part of the basal ganglia, the striatum borders the septal nuclei. The nucleus accumbens is located here as part of the ventral striatum. It has intensive fiber connections to the limbic system and is of decisive importance for psychomotoric behavior triggered by emotional stimuli. In particular, the nucleus accumbens forms the major part of our endogenous reward system, because the nucleus is intensively stimulated by unexpectedly good performance (e.g. profits, but also by addictive drugs). This is mediated by the mesolimbic connections of the mesencephalic ventral area tegmentalis (VTA) with the nucleus accumbens. The neurotransmitter dopamine is released from the axons of this tract, which, in a positive feedback, amplifies the behavior that causes a feeling of reward. Therefore, drugs are addictive, if they motivate to permanently repeat the behavior leading to the inflow of dopamine into the nucleus accumbens by circumventing the neocortex as "rational decision maker".

**Nucleus basalis (Meynert)** It is one of the most important acetylcholine-producing nuclei in the basal forebrain (located between the septum and the amygdala) and has pronounced connections to the limbic system and the neocortex.

**Nucleus caudatus** see Corpus striatum

**Nucleus dentatus** This collection of neurons forms a serrated band in the white matter of each cerebellar hemisphere. It is innervated by inhibitory neurons of the cerebellar cortex, the Purkinje cells, and by pontine cerebellar afferent axons. Its efferents project to the contralateral red nucleus and to the ventral thalamus, from which the cortical motor areas are controlled.

**Nucleus dorsalis nervi vagi** The vagal nucleus is located in the medulla oblongata medial to the nucleus nervi hypoglossi, the origin of the twelfth cranial nerve responsible for tongue movement. The visceromotor vagal nucleus is the origin of the parasympathetic, autonomic innervation of the chest and abdominal organs.

**Nucleus pedunculopontinus** Cholinergic nucleus located below the substantia nigra, which plays an important role in the ascending reticular arousal system.

**Nucleus subthalamicus** This nucleus belongs to the diencephalon in terms of its developmental history. However, due to its importance in the extrapyramidal motor system, it is functionally classified as one of the basal ganglia. Located medial to the internal capsule, the subthalamic nucleus is synaptically connected to the pallidum and activates the inhibitory, inner part of the pallidum and thus primarily inhibits motor impulses.

**Nucleus tuberomamillaris** Important nucleus in the hypothalamus, located between the tuber cinereum and the corpus mamillare. It is the only nucleus of the CNS that synthesizes the neurotransmitter histamine and thereby increases attention.

**Oligodendrocyte** see Glia

**Organoid** The few-millimeter-sized structures formed in the cell culture dish resemble the organs of the body formed from the same stem cells. Under well-defined culture conditions, organoids can therefore be cultivated from embryonic, induced or pluripotent stem cells. However, they do not contain blood vessels and are therefore only partially informative in terms of their biological relevance for neurodegenerative diseases.

**Oxidative stress** A metabolic state that results in high concentrations of reactive oxygen compounds (ROS, reactive oxygen species). In other words, it is an imbalance between oxidizing and reducing molecules. If a cell's normal repair and detoxification functions are overwhelmed, oxidative stress and, consequently, damage to lipids, proteins, and DNA may result.

**Phagocytosis** Cellular 'eating' (from the Greek, phagein), refers to the active uptake of particles or smaller cells by another cell, a special form of endocytosis.

**Phylogeny** This term refers to the evolutionary development of all living things. Phylogeny is contrasted with ontogeny, the individual development of a particular organism.

**Plasma membrane** Also known as the cell membrane, i.e. the boundary of a cell to the outside. It consists essentially of two rows of lipids, the lipid bilayer, and a multitude of membrane proteins and receptors embedded therein.

**Positron emission tomography (PET)** PET is an imaging technique in nuclear medicine that produces cross-sectional images using weakly radioactive substances. It can be used to image biochemical and physiological functions in living beings (functional imaging). The method is usually performed in conjunction with a computerized tomography (CT) or magnetic resonance imaging (MRI) as a hybrid procedure.

**Prions** These are proteins that can exist in both a physiological (normal) and a pathological (disease-causing) form. Pathogenic prions force the same proteins that are in normal conformation to adopt the pathological three-dimensional structure. This would make them quasi-infectious. Their existence was predicted as early as 1982 by Stanley Prusiner, who was awarded the Nobel Prize for this discovery in 1997.

**Progenitor cells** Cells that arise from multipotent stem cells are called progenitor cells. Since they are already committed to their future role in a particular organ, the term 'determined stem cells' also applies to them.

**Proteasome** A protein complex in the cytoplasm and cell nucleus that is of great importance for the controlled degradation of intracellular proteins. For this purpose, the proteins to be disposed of are marked by coupling of a small peptide, ubiquitin, unfolded and introduced into the proteasome. There, catalytically active subunits of the proteasome (proteolytic "scissors") cleave the protein into numerous shorter peptides, the amino acids of which are re-used.

**Pyramidal cells** These are the largest neurons in the CNS. They have long axons and complex dendrites. In the histological section, the cell body is triangular (hence the name). They occur in the cortex cerebri, hippocampus and amygdala of mammals. In the neocortex, they are primarily located in lamina III and V. Inhibitory synaptic contacts are located on the perikaryon, while excitatory synapses are found on the thorn-like extensions of the dendrites, the *spines*. The axons of the pyramidal tract, which connect the cortex with the spinal cord, can be over one meter long.

**REM sleep** The term is derived from rapid eye movements during sleep with closed eyelids. This is accompanied by a reduced tone of the skeletal muscles, increased blood pressure and specific activity patterns in the electroencephalogram (EEG). REM sleep is also referred to as paradoxical sleep or dream sleep.

**Retromer** Protein complex that regulates the transport of membrane proteins between endosomes and the Golgi apparatus. More than 100 endosomal proteins are sorted by the retromer, either towards the lysosome or back to the plasma membrane via the trans-Golgi network.

**Small interfering RNAs (siRNAs)** RNAi (RNA interference) is a cellular mechanism that cleaves double-stranded RNA (dsRNA) into several fragments of 19–23 nucleotides in length (siRNAs) using the enzyme Dicer, which are then incorporated into the RISC complex (RNA-induced silencing complex). RISC binds together with the siRNAs to the complementary DNA sequence and deactivates it. If siRNAs are added exogenously, they bind to complementary, single-stranded RNAs and thus block their normal function.

**Stem cells** These are body cells that are still in immature stage, i.e. they can differentiate into different cell types or tissues. Embryonic stem cells therefore may develop into any tissue type, whereas adult stem cells are already fixed to develop into a certain tissue. Induced pluripotent stem cells (iPS cells) are created through artificial re-programming, i.e. via external stimulation of the expression of specific genes (transcription factors). The use of iPS cells avoids ethical problems that arise when using embryonic cells from aborted fetuses in biomedical research.

**Striatum**  see Corpus striatum

**Substantia nigra**  Nucleus in the midbrain (mesencephalon) that is darkly colored due to a high content of neuromelanin (Latin, black). The densely packed, dopamine-producing neurons are located in the pars compacta. In front of it lies the pars reticulata, whose smaller, more loosely (net-like) distributed nerve cells have a high iron content. The afferents (incoming axons) to the substantia nigra originate from the motor cortex (fibrae corticonigrales) and from the caudate nucleus and the putamen (fibrae strionigrales). The efferents of the substantia nigra project as fibrae nigrostriatales to the striatum and to the thalamus.

**Subthalamus**  see Nucleus subthalamicus

**Sulcus centralis**  The central sulcus separates the frontal lobe from the parietal lobe. The precentral gyrus (primary somatomotor cortex) and the postcentral gyrus (primary somatosensory cortex) in the parietal lobe border the central sulcus.

**Sulcus lateralis**  The frontal, parietal and temporal lobes come together at this lateral furrow of the brain. By opening this sulcus the insula becomes visible.

**Synapse**  The contact points at which neurons are interconnected are called synapses. Excitation (an action potential) is transmitted via chemical messenger substances (neurotransmitters) at synapses. It is estimated that there are about 1000 trillion synapses in the adult human brain. A single nerve cell can form up to 2 million synaptic contacts.

**Synucleinopathy**  A generic term for various neurodegenerative diseases in which intracellular deposits of insoluble, pathological α-synuclein have been detected. In addition to Parkinson's disease, these include Lewy body dementia and multiple system atrophy (MSA).

**Tau**  A protein that binds to microtubules and influences their assembly. Neurodegenerative diseases associated with deposits of tau protein are referred to as tauopathies. Tau is encoded by the MAPT gene, occurs in various isoforms and can form filamentous structures.

**Tegmentum**  Part of the brainstem that is located in front of the fourth ventricle and aquaeduct (ventral tegmentum versus dorsal tectum). Numerous cranial nerve nuclei, a loose association of groups of nerve cells (the formatio reticularis), the substantia nigra, the locus coeruleus and other nuclei are located in the tegmentum.

**Telencephalon**  s. Cerebrum

**Telomer**  Telomer is the name given to the repetitive DNA at the ends of chromosomes located in the cell nucleus. Together with some associated proteins, telomeres stabilize the chromosomal endings.

**Thalamus**  see Diencephalon

**Transcription**  The production of RNA from a DNA template is called transcription. Three main groups of RNA can be distinguished: mRNA (messenger RNA) for protein biosynthesis, tRNA (transfer RNA) for coupling amino acids to a new protein synthesized by the ribosome, and rRNA (ribosomal RNA) for the construction of ribosomes. In transcription, the nucleic acid bases of DNA (e.g. the sequence T – A – C – G) are re-written into the bases of RNA (in this case A – U – G – C). Uracil is therefore incorporated instead of thymine and ribose is used in RNA instead of deoxyribose in DNA.

**Translation** The translation of the base sequence of mRNA into the amino acid sequence of a protein is called translation. The polypeptide chain to be formed consists of a total of 20 amino acids, each of which is bound by tRNA molecules in the cytoplasm and transported to the ribosome.

**Truncus cerebri** The brainstem consisting of midbrain, pons and medulla oblongata. The cerebellum is usually not considered a part of the brainstem.

**Ubiquitin-Proteasome System** see Proteasome

**Unfolded protein response (UPR)** see ERAD complex

**Ventriculi** The cavities inside the brain (ventricles) contain about 150 ml of cerebrospinal fluid (CSF), which can pass through several openings from the ventricular system into the spaces outside the brain. At the bottom of the lateral ventricles, at the roof of the third ventricle and in the area of the fourth the highly vascularized choroid plexus is found, through which the CSF is produced from serum and secreted into the ventricles (about 500 ml per day). The CSF is drained through protrusions of the meninges into venous sinuses and into the lymphatic space surrounding the cranial and spinal nerves.